Advances in Mutagenesis Research 1

Advances in Mutagenesis Research 1

Edited by G. Obe

With Contributions by

M. Bauchinger F.K. Ennever H. Hayatsu
R. Huber C. Kessler I. Mellon H. Nöthel
C.A. Smith R.M. Speed

With 63 Figures

Springer-Verlag Berlin Heidelberg New York
London Paris Tokyo Hong Kong

Professor Dr. GÜNTER OBE
FB9 der Universität
Gesamthochschule Essen
Universitätsstraße 5
Postfach 10 37 64
4300 Essen 1, FRG

ISBN-13:978-3-642-74957-5 e-ISBN-13:978-3-642-74955-1
DOI: 10.1007/978-3-642-74955-1

Softcover reprint of the hardcover 1st edition 1990

2131/3145-543210 – Printed on acid-free paper

Foreword to the Series

Mutations are permanent changes in the genetic material. These changes can comprise single genes (gene mutations), the structure of the chromosomes (chromosome mutations), or the number of the chromosomes (genome mutations). Since H.J. Muller presented his paper *The problem of genic modification* at the 5th International Congress of Genetics in Berlin on the 15th of September, 1927, in which he brilliantly showed that X-rays induce mutations in the fruit-fly *Drosophila*, we have learnt that a plethora of agents, including ionizing and nonionizing radiations, chemicals, and viruses, can induce mutations. In most of the cases, induced mutations are deleterious to the cells or the organisms in which they occur, and we cannot justify damaging the genetic material of organisms, including ourselves, by introducing man-made mutagenic agents into the environment. To prevent this, chemicals must be tested for their possible mutagenicity in a variety of test systems before they can be used. This has opened a field of applied genetic research, namely, genetic toxicology. Comparative analyses led to the concept that mutagenic agents can be expected to be also carcinogenic. The theory of the origin of cancer by mutations has gained experimental proof by the finding that oncogenes, when changed by mutations, can give rise to cancer.

Basic research in the field of mutation research has unraveled some of the molecular mechanisms underlying the origin of mutations and the complex reaction of cells to induced changes in their DNA. These cellular reactions can eventually lead to the restoration of the original structure of the DNA, but, via misrepair, can also give rise to mutations.

There are still many open questions. The molecular mechanisms leading to mutations are only partially known. In view of the fact that about 6 in 1000 newborn children have a chromosomal alteration, it would be especially important to understand how chromosome and genome mutations are produced.

Molecular changes in the DNA and the reaction of the cells to such changes result in typical mutation rates which reflect the evolutionary history of the organisms in question. Mutations are one of the sources of variability which is the prerequisite for natural selection and for evolution; but since mutations can also result in various deleterious effects, such as hereditary diseases, a population can only survive when the mutation rates are not too high and not too low, i.e., mutation rates are delicately balanced. Elevations of the mutation rates would have considerable consequences. It would lead to an increase in the frequencies of cancers and would represent a great risk for the evolutionary future of a species; a scenario in which humans are fully included.

In view of these implications, mutation research has two aims:

1. To understand the molecular mechanisms leading to mutations and
2. to prevent a thoughtless introduction of mutagenic agents into our environment.

Both aspects, namely basic and applied ones, will be treated in the new series *Advances in Mutagenesis Research*. The articles will deal with current developments in the field of mutation research and will help the reader to orient himself in this centrally important area of biology.

Prof. Dr. GÜNTER OBE

Contents

Blue Cotton – Broad Possibility in Assessing Mutagens/Carcinogens in the Environment

H. HAYATSU

Contents

Abbreviations

Trp-P-1, 3-amino-1,4-dimethyl-5*H*-pyrido[4,3-*b*]indole; Trp-P-2, 3-amino-1-methyl-5*H*-pyrido[4,3-*b*]indole; Glu-P-1, 2-amino-6-methyldipyrido[1,2-*a*:3',2'-*d*]imidazole; Glu-P-2, 2-aminodipyrido[1,2-*a*:3',2'-*d*]imidazole; AαC, 2-amino-9*H*-pyrido[2,3-*b*]indole: MeAαC, 2-amino-3-methyl-9*H*-pyrido[2,3-*b*]indole; IQ, 2-amino-3-methylimidazo[4,5-*f*]quinoline: MeIQ, 2-amino-3,4-dimethylimidazo[4,5-*f*]quinoline; MeIQx, 2-amino-3,8-dimethylimidazo[4,5-*f*]quinoxaline; 4,8-Me_2IQx, 2-amino-3,4,8-trimethylimidazo[4,5-*f*]quinoxaline; 7,8-Me_2IQx, 2-amino-3,7,8-trimethylimidazo[4,5-*f*]quinoxaline; PhIP, 2-amino-1-methyl-6-phenylimidazo[4,5-*b*]pyridine; MNU, *N*-methyl-*N*-nitrosourea.

1 Introduction

The scope of studies on mutagens widened greatly when the notion that carcinogens overlap with mutagens was introduced by Ames et al. (1973). The mutagen detection has become increasingly important since the discovery of highly potent

[1] Faculty of Pharmaceutical Sciences, Okayama University, Tsushima, Okayama 700, Japan

mutagens in cooked foods and subsequent demonstration of their carcinogenicity in rodents (Sugimura 1988). It seems now that our environment contains a great number of mutagens, and humans live surrounded by mutagens/carcinogens (Sugimura 1982; Ames 1983; Ames et al. 1987). As a result, the importance of assessing the environmental mutagens both qualitatively and quantitatively has become realized by researchers worldwide, and great efforts have been made to accomplish this task. A difficulty one encounters in attempting this assessment is that mutagens, either those of known structures or of unknown nature, are present only in tiny amounts in the environmental materials. Furthermore, these materials are almost infinite in their kinds and numbers. Consequently, methodological advancement has been, and is, very important for the development of this area of science. An efficient means to detect mutagens in complex mixtures would undoubtedly facilitate the progress of the study.

The blue cotton method is one such means, recently introduced by our own research group. This method has been used extensively for isolating mutagens and for detecting mutagenicity in the environment. In this chapter, I would like to summarize the present status of this new methodology.

1.1 What is Blue Cotton?

Blue cotton is cotton bearing covalently linked blue pigment, copper pathalocyanine trisulfonate (Hayatsu et al. 1983a). The linkage connecting the pigment to cellulose is illustrated in Fig. 1. Figure 2 shows the picture of blue cotton, together with that of blue rayon, which is a recently developed, modified version of blue cotton. The synthesis of blue cotton can be done by using a simple one-step reaction shown in Fig. 3.

The usefulness of blue cotton lies in its unique property to adsorb aromatic compounds having three or greater numbers of fused rings. The adsorption capacity is strong, and the selectivity for adsorbing this class of compound is high. The fact that many of such polycyclic aromatic compounds in the environment are mutagenic and often carcinogenic makes blue cotton extensively useful as a means to efficiently purify those mutagens from crude samples.

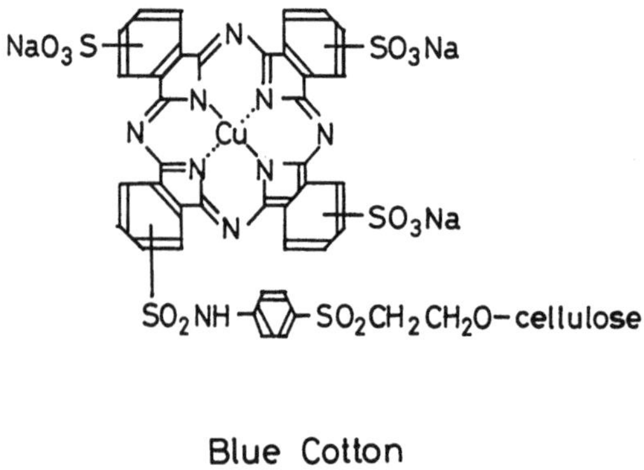

Fig. 1. Structure of blue cotton

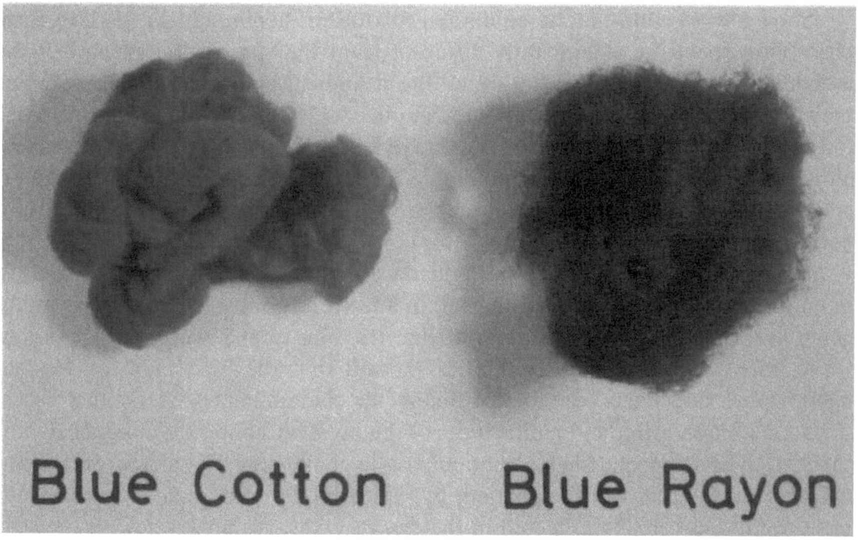

Fig. 2. Blue cotton and blue rayon

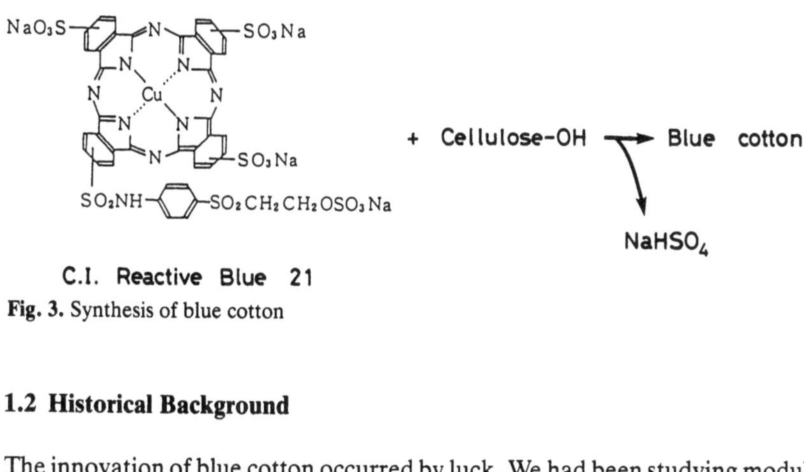

C.I. Reactive Blue 21

Fig. 3. Synthesis of blue cotton

1.2 Historical Background

The innovation of blue cotton occurred by luck. We had been studying modulators of mutagenesis by using the assay on *Salmonella* (Ames et al. 1975). Our first finding in this study was the ability of hemin to inhibit the mutagenicity of polycyclic aromatic compounds (Arimoto et al. 1980a, 1980b). Thus, the mutagenicity of Trp-P-1 in *S. typhimurium* TA98 in the presence of microsomal activating enzymes is 50% inhibited by addition of only two equivalents of hemin (Arimoto et al. 1980a). Similarly, the mutagenicity of 8 nmol benzo(a)pyrene in TA98 is 50% suppressed by the presence of 10 nmol hemin (Arimoto et al. 1980b). On the other hand, hemin does not inhibit the mutagenicity of MNU, 4-nitroquinoline 1-oxide, or several other compounds that do not possess structures of three rings or more (Arimoto et al. 1980b).

Soon it was found that an equimolar mixture of hemin and Trp-P-1 gives an absorption spectrum significantly different from the sum of the spectra of the individual compounds. We interpreted this as indicating formation of a complex between these two planar molecules (Arimoto et al. 1980a). Whether this complex formation is the cause of the mutagenicity suppression is not yet established, although it is strongly suspected to be so. Nevertheless, the fact that some interactions can occur between a polycyclic aromatic mutagen and hemin impressed me strongly.

Being in the teaching profession, I had wanted to examine whether chalks are mutagenic. The chalk powder collected in a lecture room was found to be mutagenic in the Ames test. It turned out that the blue chalks were mutagenic but white ones were not. Further studies showed that the blue pigment used to manufacture the chalks contained a mutagenic contaminant. After some discussions took place with the manufacturer of the pigment about the possible significance of the presence of such a contaminant in chalks, the manufacturer decided to improve the quality of the pigment by removing the contaminant. As a result, most of the blue chalks collected in the city of Okayama after a few years were no longer mutagenic (Hayatsu et al. 1983b).

During this investigation, the chemical nature of the blue pigment was a focus in discussion. The pigment was copper phthalocyanine (Fig. 4) . The structural similarity to hemin was obvious. At that time we already knew that hemin can form complexes with polycyclic compounds; therefore, it was a straightforward guess that copper phthalocyanine might behave like hemin towards polycyclic aromatic compounds. An interesting question emerged whether these molecules can serve as a ligand to adsorb polycyclic aromatic compounds. To explore this possibility, we wanted to have these molecules connected to some carrier. Our attempt at that time to covalently link hemin to a solid support had failed. Therefore, we decided to investigate the use of copper phthalocyanine or its derivative as a ligand on a solid support for adsorbing polycyclic mutagens. As the support, absorbent cotton appeared to be an attractive candidate because it swells in water to allow an efficient contact with molecules in solution, and because the manipulation is expected to be easy. Fortunately, there was a copper phthal-

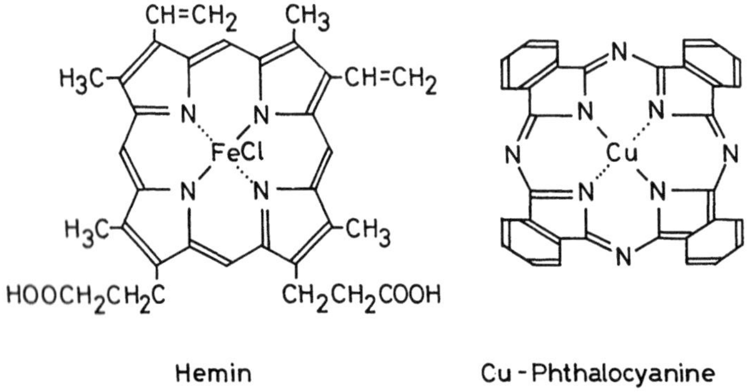

Fig. 4. Structures of hemin and copper phthalocyanine

ocyanine derivative perfectly suited for our purpose. C.I. Reactive Blue 21 can react with hydroxyl groups to form a stable ether linkage. It bears three sulfonate substituents on the ring, so that it has a strong hydrophilic character, a property that seemed important for a ligand intended to be used in aqueous media. As is described below, blue cotton prepared from absorbent cotton and C.I. Reactive Blue 21 has proved to be an excellent adsorbent for polycyclic aromatic compounds.

2 Preparation and Properties of Blue Cotton

Phthalocyanines are the representative of blue and green pigments that have commercial values (Moser and Thomas 1983). Many of blue-pigmented materials we see everyday are colored with phthalocyanines. There are a large number of phthalocyanine derivatives potentially useful for the purpose of our present interest, namely, as an adsorbent of polycyclic compounds. In addition, the support to carry the phthalocyanine ligand can be anything from cellulose to glass beads. Therefore, the combination, C.I. Reactive Blue 21 plus cotton, is only one of the various possibilities.

2.1 Preparation

As the scheme in Fig. 3 shows, the reaction between C.I. Reactive Blue 21 and cellulose is of a Schotten-Baumann type. The condensation proceeds in an aqueous medium under a sodium carbonate-alkaline condition at 70 °C (Hayatsu et al. 1983a). The whole process requires about 2 h, during which period the absorbent cotton is gently swirled in the reaction flask. The stained cotton is collected on a filter and washed with water and then with dimethylsulfoxide. The dimethylsulfoxide washing is a necessary step to efficiently remove unbound pigment from the cotton. Generally, several repeated soakings of the cotton in dimethylsulfoxide are required to obtain a very faint blue washing. To ensure removal of any polycyclic aromatic compounds that may have been adsorbed to the blue cotton during these processes, the cotton is then washed with a mixture of methanol and concentrated ammonia (50:1, v/v) and finally with methanol.

The dried sample may be analyzed by quantitating the copper content by atomic absorption. A simpler way of its analysis is currently used in our laboratory (unpublished work). It consists of decomposing the cellulose with 60% perchloric acid by heating at 90 °C for 30 s, making the solution alkaline by diluting with an excess NaOH, and finally subjecting the solution to colorimetric determination of the blue pigment. For this spectroscopic determination, a standard curve is provided by submitting C.I. Reactive Blue to the same treatment. The values thus obtained by this method are consistent with those found by the copper analysis.

The blue cotton samples prepared under conditions specified in the literature (Hayatsu et al. 1983a) have shown reproducible copper phthalocyanine contents of ~10 μmol per g dried material. Although the phthalocyanine content in the cotton can be increased by submitting the stained cotton again to the treatment with C.I. Reactive Blue 21, we thought that the once-stained material was satisfactory for use in mutagen studies.

An obvious extension of this synthesis is to explore the variation of the kind of cellulose employed as the carrier of the ligand. For example, cellulose powder can be stained similarly, and the copper phthalocyanine content of the blue cellulose powder prepared is approximately the same as that of blue cotton (see Sect. 6). An interesting observation was recently made when we used a special kind of rayon. This rayon has been obtained from Daiwabo Co. (Kakogawa, Hyogo), a material called by the company "amorphous rayon staple fibers". This fiber lacks the rigid filamentous structure that can be found in regular rayon fibers, and consequently provides a better access for reagent molecules to its surface. On staining this rayon fiber with C.I. Reactive Blue 21 using the conditions described above, about three times the amount of pigment was bound to the cellulose, as compared to the pigment content of blue cotton (see Fig. 2).

2.2 Properties

Phthalocyanines show outstanding stability to light and heat (Moser and Thomas 1983). The pigment content of blue cotton, as determined by the perchloric acid solubilization, did not change after allowing the cotton to stand on a laboratory bench in an opened container for 6 months without particular cautions for light. A gradual reduction of the color occurs when blue cotton is exposed to a continuous flow of tap water. This decoloration is apparently due to chlorination of the pigment: treatment of blue cotton with water containing 10 ppm Cl_2 results in a rapidly diminishing blue.

Because of its strong capacity to adsorb polycyclic aromatics, it is advisable that old samples of blue cotton be refreshed before use by washing with methanol-ammonia and methanol.

Blue cotton contains a small amount of nonbonded pigment that has remained in the material even after an exhaustive washing. The blue pigment eluted with methanol-ammonia from a blue cotton preparation corresponded to 6 ppm of the cotton.

Blue rayon prepared from amorphous rayon fibers contains 30 μmol copper phthalocyanine per gram. Because of its particular physical property, this fiber can be efficiently washed with solvents. Free blue pigment that has been held in the rayon is less than 1 ppm. Amorphous rayon fiber is a little weaker than regular rayon fiber against tensile forces: the amorphous rayon has a tensile strength of 1.97 g/denier, whereas regular rayons have the strength of 2.5–3.1 g/denier and cotton that of 3.0–4.9 g/denier (Handbook for Chemical Fibers 1985). Therefore, if a suspension of blue rayon in water is shaken strongly, a portion of the rayon may be split to form short pieces of fibers. In our hands, however, these fines are generally less than 0.5% in weight of the rayon placed in the solvent. When necessary, the fines may be collected by use of a gauze.

3 Specificity as a Mutagen Adsorbent

3.1 Classification of Adsorbable and Nonadsorbable Compounds

Various mutagens dissolved in saline can be adsorbed to blue cotton and can be recovered from the cotton by elution with an organic solvent (Hayatsu et al. 1983a). The adsorption is achieved simply by a batch process: shaking the mutagen solution with the blue cotton added. Elution with a mixture of methanol and a small amount of concentrated ammonia (usually used at a ratio of 50:1, but sometimes at a ratio of 1000:1) generally gives excellent recoveries of adsorbed mutagens. Table 1 shows results for several typical compounds, and Fig. 5 is a summary of experiments for 62 compounds. Compounds with three or more fused rings in their structures are adsorbed efficiently to blue cotton and can be recovered satisfactorily, whereas compounds with 2-, 1- or 0-rings are adsorbed only to a small extent. There are several exceptional cases. 1,N^6-ethenoadenosine (No. 17) is unadsorbed to blue cotton, although it has a 3-ring system. 1,N^6-ethenoadenine, the base portion of No. 17, is also unadsorbable (data not shown). The reason for this phenomenon is not clear; this 3-ring system may have some structural anomaly (Scheller and Sigel 1983). PhIP (No. 50) and quercetin (No. 51) appear also to be exceptional; however, they have a 2-ring and a 1-ring directly linked and conjugated, and therefore would possess coplanar sizes comparable to

Table 1. Adsorption and recovery of compounds by blue cotton[a]

Compound	Adsorption (%) Blue cotton	Adsorption (%) Plain cotton	Overall recovery by blue cotton (%)
Glu-P-1	85	9	80
MeIQx	94	12	94
Daunorubicin	99	22	65
Aflatoxin B_1	84	26	63
4-Nitroquinoline l-oxide	11	0	Not done
p-Nitrophenol. Na	5	0	Not done
[^{3}H]Histidine	–	Not done	0.06
[^{14}C]Nitrosodimethylamine	0	Not done	Not done

[a] A 5-ml solution of a compound in 0.15 M NaCl (concentrations, 0.1 – 100 μM) was treated with blue cotton (50 mg) by shaking at room temperature for 30 min. The cotton was taken up and squeezed to combine the cotton-held portion of the solution to the mother liquor. A fresh batch of blue cotton (50 mg) was added, and the treatment was repeated. The adsorption extent was determined by quantifying the compound that remained in the solution (either by spectrometry, fluorometry, or radioactivity measurement). The two 50-mg cotton samples were combined, wiped with a paper towel, washed twice with 0.15 M NaCl, freed of moisture by use of a paper towel, and then eluted with 5-ml methanol-concentrated ammonia (50:1) during gentle shaking for 30 min at room temperature. This elution was repeated once more, and the combined eluant was evaporated under reduced pressure. The residue was subjected to quantification of the compound. For the histidine adsorption, 1 mM [^{3}H]histidine at 230000 dpm was used, and the radioactivity recovered was 146 dpm, corresponding to 0.06 % overall recovery.

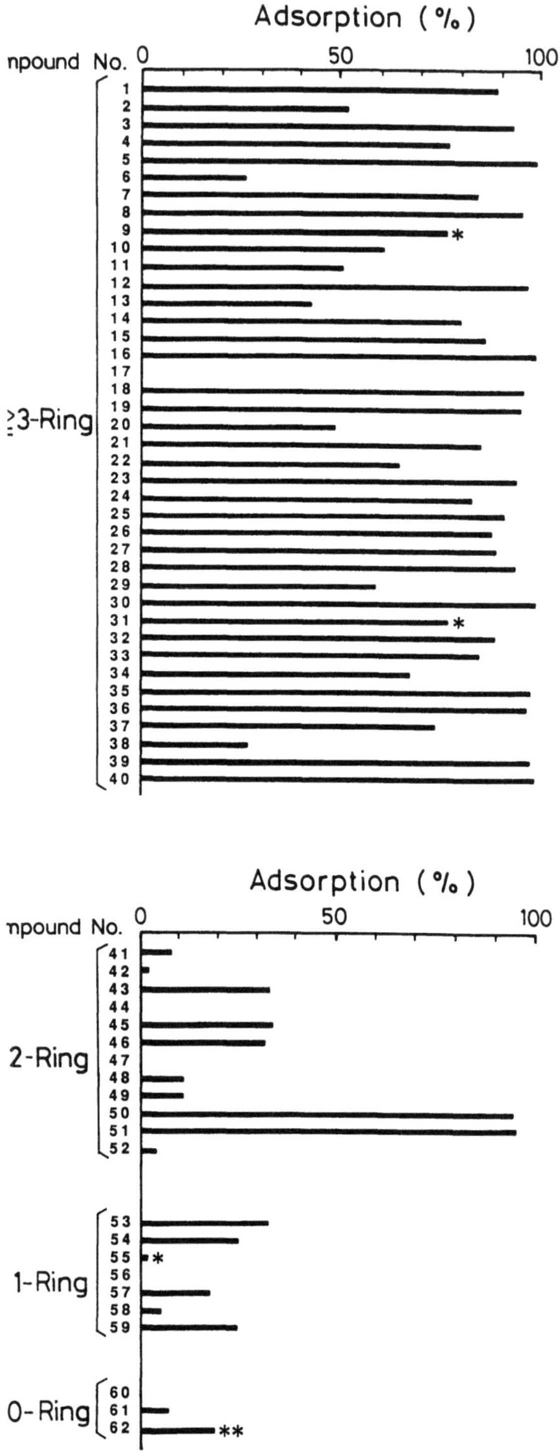

Fig. 5. Adsorption of 62 compounds to blue cotton. The experimental procedure is described in Table 1. The *star* (*) indicates overall recovery, rather than the extent of adsorption; ** the experiment for fecapentaene-12 was done with blue rayon: 20 mg rayon for 43 nmol fecapentaene-12 in 5 ml saline under an argon atmosphere, and the fecapentaene-12 was quantified spectrophotometrically. The fecapentaene-12 used was a generous gift of Dr. K. Wakabayashi of the National Cancer Center Research Institute, Tokyo, who prepared the sample as reported (Shioya et al. 1989). Compound numbers and names are as follows: *1* AαC; *2* 2-acetylaminofluorene; *3* N-acetyl Trp-P-1; *4* N-acetyl Trp-P-2; *5* acriflavine hydrochloride; *6* actinomycin D; *7* aflatoxin B_1; *8* 9-aminoacridine; *9* 2-aminoanthracene; *10* bellidifolin (courtesy of Dr. H. Kanamori, Hiroshima Institute of Hygiene); *11* benz(a)anthracene 5-methylenesulfate (courtesy of Prof. T. Watabe of Tokyo College of Pharmacy); *12* benzo(a)pyrene; *13* carminic acid; *14* chlorpromazine hydrochloride; *15* copper phthalocyanine tetrasulfonic acid; *16* daunorubicin; *17* 1,N^6-ethenoadenosine; *18* hetridium bromide; *19* ethyl eosin; *20* fluorescein; *21* Glu-P-1; *22* Glu-P-2; *23* harman; *24* hemin; *25* IQ; *26* MeAαC; *27* MeIQ; *28* MeIQx; *29* 8-methoxypsoralen; *30* methylene blue; *31* 1-nitropyrene; *32* norhaman; *33* phenazine A (courtesy of Prof. T. Okuda of Okayama University) (Okuda et al. 1981); *34* phenazine C (Okuda et al. 1981); *35* quinacrine hydrochloride; *36* rhodamine 6G; *37* riboflavine; *38* swertianolin (courtesy of Dr. H. Kanamori); *39* Trp-P-1; *40* Trp-P-2; *41* adenine; *42* adenosine; *43* Amaranth *44* ATP; *45* Carbadox; *46* 6-dimethylallylaminopurine; *47* NADH; *48* naphtalene 1-methylenesulfate; *49* 4-nitroquinoline 1-oxide; *50* PhIP; *51* quercetin; *52* tryptophan; *53* 4-aminobiphenyl; *54* furylfuramide; *55* histidine; *56* mitomycin C; *57* 4-nitro-*o*-phenylenediamine; *58* *p*-nitrophenol.Na; *59* phenolphthalein; *60* nitrosodimethylamine; *61* oleic acid.Na; *62* fecapentaene-12

a 3-ring system. Fecapentaene (No. 62) seems to have a weak affinity to the blue dye. The linear but planar structure of this aliphatic compound may contribute to the affinity found. Control experiments using plain absorbent cotton (Table 1) clearly shows that the adsorption to blue cotton is dependent on the pigment bound to cotton. A cotton sample, which has undergone the sodium carbonate treatment in the absence of the dye, is also tested as a control. There was no significant difference found between the carbonate-treated and the plain cotton samples in their adsorption of compounds. The effect of pH of the mutagen solution on the adsorption was studied for Trp-P-2, and the adsorption was equally efficient in the pH range 4–10. It was also demonstrated that [^{3}H]Trp-P-2 and [^{14}C]acetylaminofluorene can be similarly recovered by blue cotton from solutions of these compounds in human urine or bovine serum.

Most of the mutagens that have been found to adsorb to blue cotton are those that give positive mutagenicity in the *Salmonella*/microsome test using the strain TA98 in the presence of S9 mix. The compounds that can interfere with the Ames test, i.e., histidine and oleic acid (Hayatsu et al. 1981), are practically nonadsorbable to blue cotton.

Since the ligand has sulfonate groups, blue cotton is expected to have a cation-exchanging capacity. However, the adsorptions in experiments shown in Fig. 5 took place in 0.15 M NaCl solutions, and therefore such ionic affinities would have contributed little to the observed adsorptions. In fact, many of the adsorbable compounds do not have dissociable groups, e.g., acetylaminofluorene and benzo(a)pyrene. It should be borne in mind, however, that blue cotton has this ionic property and that such a property may become manifest in solutions of low salt concentrations.

Our experimental results, together with those reported from other laboratories (see Sect. 4), indicate that blue cotton has a selective affinity to compounds having three or more fused rings. These compounds are commonly planar in their molecular forms. Consequently, it is conceivable that they can form face-to-face hydrophobic complexes with copper phthalocyanine moiety, which has a large planar surface in the molecule.

As expected, blue rayon shows similar adsorption chararcteristics. Blue rayon has a higher capacity and a lower free-pigment content than blue cotton, as already discussed in Sect. 2. Although its mechanistic strength is somewhat lower than blue cotton, it does not seem to create a serious problem: for example, blue rayon did not show any weight loss when allowed to hung in water of a river for 2 days (Sakamoto and Hayatsu 1988).

Table 2 shows results of a series of experiments in which (1) the efficiency of adsorption was compared between blue rayon and blue cotton, and (2) the efficiency of elution was compared between methanol-only and methanol-ammonia for various compounds on blue-rayon and -cotton. (1) With equal masses of the rayon and cotton, the adsorption extents are generally greater with the rayon (note that the molar amounts of blue pigment used in these experiments are different between the rayon and the cotton; the molar amounts of pigment in the cotton are threefold less than those in the rayon, and are in some cases less than the stoichiometric amount from the compound used). (2) Many of the compounds

Table 2. Comparison of blue rayon and blue cotton in their adsorption capacities, and the efficiency of eluting solvents[a]

Compound	μmol	Adsorption (%)		Overall recovery (%)			
		Rayon	Cotton	Elution with methanol		Elution with methanol-NH_3	
				Rayon	Cotton	Rayon	Cotton
Trp-P-1	0.1	95	89	10	15	94	90
Glu-P-1	0.25	61	30	44	18	52	22
Glu-P-2	0.25	45	20	34	15	36	15
MeIQx	0.1	83	59	51	43	78	57
IQ	0.04	79	58	73	60	81	72
Ethyl Eosin	0.05	66	37	52	33	52	34
2,4,5,7-Tetranitro-9-fluorenone	0.1	24	25	35	24	35	24
Quinacrine	0.5	83	33	0	4	73	34
Chlorpromazine	0.2	61	25	39	24	66	26
PhIP	0.01	–	–	69	67	75	72
4-Nitroquinoline 1-oxide	0.3	6	2	–	–	–	–
Carbadox	0.2	26	13	23	13	–	–
Tryptophan	0.3	0	0	–	–	–	–
Furylfuramide	0.2	7	0	16	0	–	–
p-Nitrophenol-Na	0.2	2	0	–	–	–	–
Benzo(a)pyrene	0.0004	96	88	63	76	68	87
					(Dioxane	87	90)

[a] A 5-ml saline solution of a compound was treated once with 20 mg of blue rayon or blue cotton for 5 min under mechanical shaking. The adsorption extent was measured spectrophotometrically. The rayon or cotton was taken out, washed with 1 ml water, and dried by use of a paper towel. The rayon or cotton was then eluted with 5 ml methanol by shaking for 5 min. The eluted compound was quantified spectrophotometrically, and then to the methanol-rayon (or-cotton) mixture was added 0.1 ml concentrated ammonia. The mixture was further shaken for 5 min and the eluted compound was quantified. For adsorption of benzo(a)pyrene, a solution in dioxane-saline (2:8) was used because of the insoluble nature of this compound in saline.

can be efficiently eluted with methanol-only, but Trp-P-1 and quinacrine were not elutable. Trp-P-1 and quinacrine can be quantitatively eluted from the cotton with methanol-ammonia. For benzo(a)pyrene, and probably for other polycyclic aromatic hydrocarbons as well, the elution by 100% dioxane appears to be the most appropriate.

3.2 Mechanism of Adsorption

At the time when we made these observations in 1983, there appeared to be no report on the affinity of aromatic compounds to copper phthalocyanines, except for the literature on the adsorption of gaseous benzene derivatives to the surface of crystalline copper phthalocyanine (Dean et al. 1978; Schröder 1982). This lack

Table 3. Analysis of complex formation between three-ring aromatic compounds and copper phthalocyanine trisulfonate covalenty linked to cellulose (blue cotton) (Hayatsu et al. 1986a)[a]

Compound	Dissociation constant for the complex formation ($k_D \times 10^6$, M)	Molar ratio at saturation (compound/Cu-pc-SO_3^-)
9-Aminoacridine	5.90	1.10
Trp-P-2	2.90	1.25
Quinacrine	4.83	1.03
2-Aminofluorene	73.1	1.00
2-Acetyaminofluorene	18.5	0.96

[a] The adsorption is in 0.05 M sodium phosphate buffer, pH 7.4, containing 0.2 M sodium chloride, at 25° C. k_D = [C] × [Cu-pc-SO_3]/[C· Cu-pc-SO_3^-], where [C] is the concentration of the compound that remained in the solution after equilibration, [Cu-pc-SO_3^-] that of the complex, which is identical to the concentration of compound adsorbed.

of information is possibly due to the fact that copper phthalocyanine sulfonates (hereafter referred to as Cu-pc-SO_3^-) form dimers and higher aggregates in aqueous solutions in a reversible manner (Blagrove and Gruen 1972, 1973). Any attempt to study quantitatively a complex formation between Cu-pc-SO_3^- and another substance would have been hampered by this complicated situation.

In blue cotton Cu-pc-SO_3^- is fixed on a solid support; hence aggregations of these Cu-pc-SO_3^- residues would be difficult. It should be noted that the procedure for preparing blue cotton involves washing the material with dimethylsulfoxide and methanol (Hayatsu et al. 1983a), solvents in which such aggregates would be dissociated (Blagrove and Gruen 1972, 1973). Furthermore, in blue cotton a complex already formed between a Cu-pc-SO_3^- moiety and a given compound is expected to exert little interference with other Cu-pc-SO_3^- moieties to make complexes.

Based on these considerations, we analyzed quantitatively the adsorption of five different 3-ring compounds to blue cotton (Hayatsu et al. 1986a). The medium used for the adsorption was 0.05 M sodium phosphate – 0.2 M NaCl, pH 7.4, and two types of titration were done. First, a fixed amount of blue cotton was titrated with a compound, and secondly, a fixed amount of a compound was titrated by adding increasing amounts of blue cotton. Both of the titrations indicated one-to-one complexing, and the results obtained by the second type of titrations are given in Table 3 (see Hayatsu et al. 1986a for details). All five compounds examined show qualitatively the same behavior, forming one-to-one complexes with Cu-pc-SO_3^-, with the dissociation constants ranging 10^{-6}–10^{-5} M.

Further evidence for the presence of molecular interactions between these adsorbable compounds and Cu-pc-SO_3^- was obtained in spectroscopic studies. Figure 6 shows visible absorption spectra of copper phthalocyanine disulfonate (spectrum 1), Trp-P-1 (spectrum 2), and their mixture (spectrum 3), each at 20 μM concentration in 50 mM sodium phosphate buffer, pH 7.0. A large shift of the peak at 606 nm to a longer wavelength is seen. Similar measurements were made with other compounds, and the results are summarized in Table 4. Com-

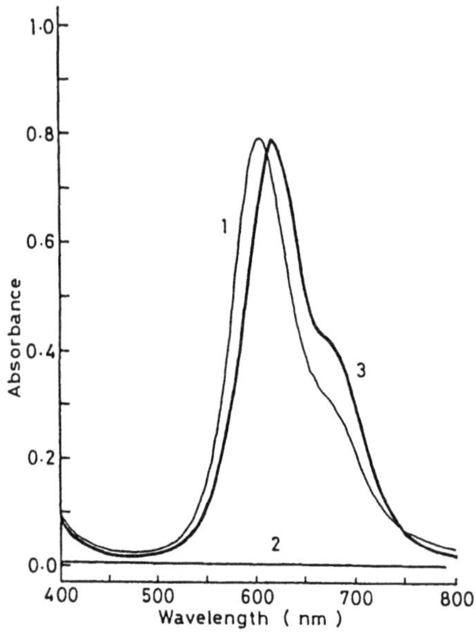

Fig. 6. Visible absorption spectra of 1–20 μM copper phthalocyanine disulfonate *2*; 20 μM Trp-P-1 and *3* 20 μM copper phthalocyanine disulfonate plus 20 μM Trp-P-1. The solutions were in 50 mM sodium phosphate buffer, pH 7.0

Table 4. Parallelism between adsorption of compounds to blue cotton and their ability to change the absorption spectrum of copper phthalocyanine disulfonate[a]

Compound	Adsorption (%)	Shift of λ_{max} at 606 nm (nm)
Trp-P-1	96	+ 12
Trp-P-2	95	+ 10
Glu-P-1	49	+ 4
Glu-P-2	28	+ 2
AαC	49	+ 2
MeAαC	72	+ 6
9-Aminoacridine	78	+ 10
Daunorubicin	82	+ 14
2-Acetylaminofluorene	20	+ 4
Ehtidium bromide	92	+ 14
Quinacrine	75	+ 18
8-Methoxypsoralen	33	+ 2
Norharman	65	+ 4
Carbadox	8	0
Quinoxaline 1,4-dioxide	11	0
4-Nitroquinoline 1-oxide	1	0
Adenine	3	0
4-Nitro-*o*-phenylenediamine	9	0
p-Nitrophenol·Na	0	0
Tryptophan	5	0
Histidine	0	0

[a] The solutions were 20 μM copper phthalocyanine disulfonate Na_2 (Cyanine Blue S-1, Sumitomo Chemical Industries, Inc.) and 20 μM test compound, dissolved in 50 mM sodium phosphate buffer, pH 7.0. The adsorption extents were determined by the procedure described in the legend to Table 1, but with a single blue-cotton (50 mg) treatment instead of a double treatment in the experiments of Table 1.

pounds that can adsorb to blue cotton induce spectral shifts, but those that do not adsorb can induce no shift in the copper phthalocyanine spectrum. It is noteworthy that there exists a tendency that strongly adsorbable compounds induce large spectral changes.

In experiments described in Table 1 and Fig. 5, no spectral changes were detected for the compounds before and after the blue cotton treatment, an indication that the pigment did not catalyze chemical transformations such as aerobic oxidation. During these treatments, no particular care was taken to protect the materials from light.

The fact that the compounds adsorbed to blue cotton can be recovered by elution with hydrophilic organic solvents indicates that the complex formation does not occur in these solvents. As already discussed, methanol-ammonia is the most effective among the eluents examined. It appears possible that ammonia helps dissociate the complex by coordinating itself to the central metal of the ligand.

4 Use of Blue Cotton in Studies of Mutagens/Carcinogens

A large number of environmental mutagens and carcinogens have polycyclic aromatic structures; polycyclic aromatic hydrocarbons and food pyrolysis products belong to this class of compounds. Micotoxins, such as aflatoxin B_1, can adsorb to blue cotton (Table 1), although its structure is not perfectly planar. It can therefore be expected that blue cotton is useful in extracting these types of

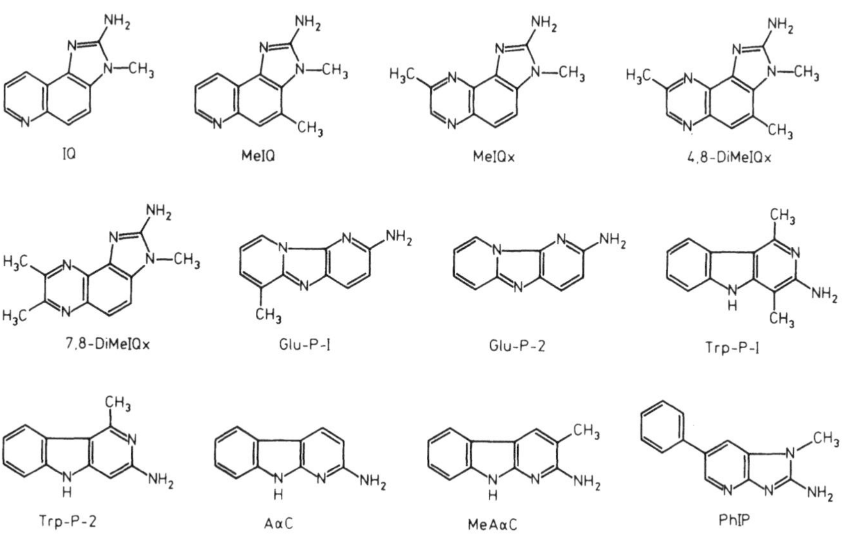

Fig. 7. List of food pyrolysate mutagens. The drawing of compounds except PhIP is taken from Sugimura (1985), and that of PhIP from Felton et al. (1986)

mutagenic compounds from complex mixtures. In this section, I would like to summarize the contents of reports on applications of blue cotton that have been published up until June 1988.

4.1 Mutagens in Food

A list of mutagenic heterocyclic amines that can be formed by pyrolysis of foods, proteins and amino acids is given in Fig. 7. Most of them are proven carcinogens in rodents (Sugimura 1988).

A typical example for the use of blue cotton to demonstrate the mutagenicity of cooked food is the following (Hayatsu et al. 1983a): a hot-water extract of pan-fried ground beef is prepared and treated with blue cotton at room temperature. The cotton is taken up, washed with distilled water to remove the beef soup, dried by a paper towel, and then eluted with methanol-ammonia. After removal of the solvent by evaporation, the residue is submitted to the Ames test on *S. typhimurium* TA98 with metabolic activation. With this simple work-up, mutagenicity of cooked beef can be demonstrated in a clear-cut manner: a linear dose response with respect to the amount of beef, and an increase of mutagenicity with the increase of cooking time.

For extraction of mutagenic compounds from Difco Bacto beef extract, the sample was first diluted by dissolving in water, and the solution was then treated with blue cotton (Hayatsu et al. 1983c). In this study of beef extract mutagens, we carried out several cycles of blue cotton adsorption, i.e., a blue cotton extract was again dissolved in water, and the solution was treated with a new batch of blue cotton. With this repetition of the cycle, an efficient concentration of the mutagenic component was achieved with only a small loss of total activity (Table 5). By further purification with HPLC, the presence of MeIQx and IQ in the beef extract was demonstrated, a result consistent with the literature reporting the presence of these mutagens in cooked beef (Kasai et al. 1981a, b).

More recently, Sugimura and co-workers applied the blue cotton adsorption in the first step of quantifying IQ and MeIQx in beef extract (Takahashi et al. 1985a). They were able to achieve 10^4-fold concentration of MeIQx-containing fraction by use of two cycles of blue cotton extraction, with an overall recovery of 71% with respect to MeIQx. By a similar procedure, 4,8-Me_2IQx in beef extract was quantified (Takahashi et al. 1985b). These workers also succeeded in detecting Trp-P-2 in blue cotton extracts of both bacteriological-grade beef extract and broiled beef (Takahashi et al. 1985c).

When one intends to detect mutagenicity in a crude sample, a necessary step is to reduce the amount of the sample to make it suitable for subjecting to an assay. For such purpose of concentrating mutagenic fractions, blue cotton offers a simple way, as exemplified in the efficient concentration of mutagens from beef extracts. Crude materials such as food often contain inhibitors for mutagens. Free unsaturated fatty acids, such as oleic acid and linoleic acid, are ubiquitously present in fats and oils, and they can mask the mutagenicity of compounds (Hayatsu et al. 1981). We have shown that oleic acid interferes with the detection

Table 5. Efficiency of repetitive blue cotton processing in purification of mutagens in beef extract (Hayatsu et al. 1983c)

	Sample	Blue cotton used	Material obtained	
			Weight (mg)	Mutagenicity[a]
Cycle 1	10 g Difco Bacto beef extract in 100 ml H_2O	0.8 g × 3[b]	15.6	150400 (100%)
Cycle 2	15.6 mg in 100 ml H_2O	0.4 g × 2	0.3	118200 (78%)
Cycle 3	0.3 mg in 50 ml H_2O	0.2 g × 2	0.1	101800 (68%)

[a] The mutagenicity was assayed for small fractions of the samples using *S. typhimurium* TA98 with metabolic activation. The numbers represent those of revertants corresponding to the total material.

[b] This expression means that the blue cotton adsorption was done three times, with 0.8 g fresh cotton at each adsorption.

of mutagenicity of various compounds in the Ames test, and have presented evidence to indicate that this inhibition occurs by entrapping the mutagens in micelles of the fatty acid (Hayatsu et al. 1983d). Since these fatty acids are only very poorly adsorbed to blue cotton, samples prepared by the blue cotton method are usually free from this problem.

With these advantages of blue cotton in mind, we started a survey of mutagenicity in processed food commercially available in the market (Kikugawa et al. 1985). A food sample was extracted with boiling water, and the aqueous solution was subjected to the blue cotton procedure to prepare a material for assays in *S. typhimurium* TA 98 and TA100, with and without metabolic activation. Soon we found that smoked, dried bonito (*katsuobushi*, in Japanese) is mutagenic (Kikugawa et al. 1985). This item is a traditional, popular food in Japan. The blue cotton extract was further fractionated by chromatographic techniques, and MeIQx was identified as a major mutagenic component (Kikugawa et al. 1986a). A minor mutagenic component was also isolated, and was suggested to be 4,8-Me_2IQx (Kikugawa et al. 1986a). Later, MeIQx and 4,8-Me_2IQx were also found in other Japanese smoked, dried fish products (Kato et al. 1986). Since raw bonito does not contain these mutagens, investigation was made to find which stage of processing is responsible for the mutagen formation. It turned out that the process called *baikan*, in which the fish is dried at 80–120°C for a period of several days, is responsible (Kikugawa et al. 1986b). Consistent with this finding, heating various fish at 100°C for 48 h, which does not result in charring of the fish, can give rise to formation of MeIQx and 4,8-Me_2IQx (Kikugawa and Kato 1987). In these studies, the blue cotton method was used extensively.

Blue cotton has since been utilized by many researchers as a regular step for preparing partially purified samples of food pyrolysate heterocyclic amines to be submitted to HPLC analysis. Thus, Edmonds et al. (1986) quantified IQ and MeIQ in broiled salmon, Aeschbacher and co-workers (Aeschbacher et al. 1987; Turesky et al. 1988) analyzed IQ and MeIQx in cooked beef products, and Manabe and co-workers (1988a) detected Glu-P-1 and Glu-P-2 in Worcestershire sauce. Recently, Zhang et al. (1988) analyzed Chinese cooked fish and meat, and detected IQ, MeIQ, MeIQx, 4,8-Me_2IQx, and PhIP.

Survey of mutagenicity in food by the blue cotton method has continued to offer information about new sources of mutagens. While boiled rice is free of mutagenicity, a broiled rice ball is positive in the Ames test: Muraoka's observation of this (1986) is important because broiled rice ball is a popular item of Japanese food. The survey which led to the finding of the *katsuobushi* mutagenicity as discussed above (Kikugawa et al. 1985) has produced further information, namely, coffee may contain some mutagens other than methyl glyoxal (Nagao et al. 1985). Further investigation on this subject has shown that MeIQ-like mutagens are present in roasted coffee beans as a tightly bound form, extractable only by methanol-ammonia (Kikugawa et al. 1988).

An important question in the food-mutagen problems is how these heterocyclic amines are formed. This question was answered by a series of work performed in Sugimura's laboratory. Thus, on heating a mixture of creatinine, *D*-glucose, and glycine, MeIQx (Jägerstad et al. 1984) and 7,8-Me_2IQx (Negishi et al. 1984a) are formed. Another mutagen, 4,8-Me_2IQx, can be formed by replacing glycine in the mixture by threonine (Negishi et al. 1984b, 1985). These model reactions have indicated the possible routes of forming these mutagens in the cooking of food. In these studies, the blue cotton procedure was used in the isolation of the products. Muramatsu and Matsushima (1985) found the formation of MeIQx and 4,8-Me_2IQx on heating mixtures which were similar in composition to those of the Sugimura group.

4.2 Cigarette Smoke, and Opium Pyrolysate

Cigarette smoke intake is associated with human cancer development (Doll 1977). An association is known between the ingestion of opium pyrolysates and the incidence of esophageal cancers in Iran (references cited in Friesen et al. 1987). Mutagenic components in these plant pyrolysis products have been studied by use of blue cotton.

Yamashita et al. (1986a, b) subjected a basic fraction of cigarette smoke condensates to the blue cotton extraction, and from the extract they isolated IQ by liquid chromatography. The mutagenicity of the IQ accounted for 1% of the activity of the basic fraction, which in turn corresponded to 42% of the activity of total cigarette condensates. The recovery of IQ during the whole process, including the blue cotton extraction and the chromatographies, was estimated to be 70% by an experiment in which an authentic IQ was added to the smoke condensate. In addition to IQ, these researchers have detected Trp-P-1, Trp-P-2,

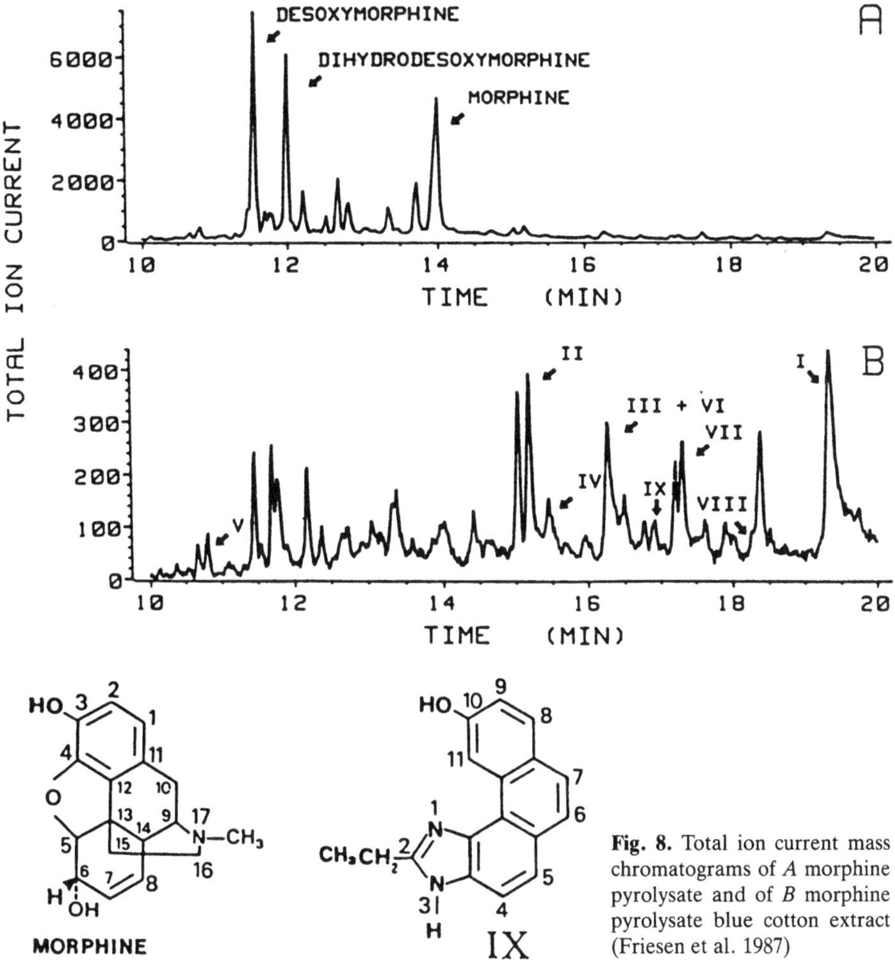

Fig. 8. Total ion current mass chromatograms of *A* morphine pyrolysate and of *B* morphine pyrolysate blue cotton extract (Friesen et al. 1987)

AαC and MeAαC in cigarette smoke condensates, again by use of blue cotton (Yamashita et al. 1986b).

A useful application of blue cotton has been made in the isolation and identification of new mutagens from opium pyrolysates (Friesen et al. 1987). In this study by scientists at the International Agency for Research on Cancer, Lyon, they took the advantage of the selective adsorbability of blue cotton for planar compounds. The prepared morphine pyrolysates were subjected to the blue cotton extraction. As Fig. 8 shows, gas chromatographic analysis of morphine pyrolysate and its blue cotton extract indicate that a selective concentration of compounds I-IX took place. These compounds were further purified and identified as substituted hydroxyphenanthrenes having strong mutagenicities in bacteria. The structure of compound IX, a compound having the strongest activity among these hydroxyphenanthrenes, is also illustrated here. It is remarkable that whereas morphine families, all retaining the morphine D-ring perpendicular to the phenanthrene ring system, are recovered only poorly by this extraction procedure, almost 100% of compounds I-IX, which are planar, are extractable.

4.3 Urines and Feces

Analysis of these excretions gives important informations about metabolisms of mutagens. The results of analysis may also provide a glimpse of the internal exposure of the human body to mutagens; there is a recent review on this subject (Venitt 1988). Human excretions, especially feces, are materials for analysis of which the availability of easy manipulation, such as the blue cotton extraction, is obviously valuable.

In our early report (Hayatsu et al. 1983a), we showed that blue cotton extracts of smoker's urine are mutagenic in the *Salmonella* test on strain TA98 with metabolic activation, a result consistent with the known mutagenicity of smoker's urine (Yamasaki and Ames 1977). The ease of manipulation has allowed a student of our group to carry out time-course studies on urinary mutagenicity during smoking and no smoking sessions over a period of several days (Kobayashi and Hayatsu 1984). It was found that mutagenicity in the urine appears and disappears rapidly in response to the start and stop of cigarette smoking. In this study, a comparison was made for the efficiency of extraction methods: the blue cotton, the XAD-2 resin (Yamasaki and Ames 1977) and the Sep-Pak C_{18} column (Becher and Bjørseth 1983). Among six individual after-smoking samples, five gave the highest responses with blue cotton and one gave the highest response with XAD-2 (Table 6). With blue cotton, mainly polycyclic mutagens must be monitored, whereas with XAD-2, organic substances in general would have been included in the assay samples (Yamasaki and Ames 1977). While the blue cotton method is biased in this sense, its high sensitivity is undoubtedly useful in a general survey of urinary mutagenicity.

Mohtashamipur et al. (1985) reported that either with the blue cotton method or with the proposed chloroform extraction procedure, a clear correlation can be found between urine mutagenicity and the number of cigarettes smoked. In contrast, in the use of XAD-2 chromatography, these researchers were unable to find a clear dose response between the number of cigarettes smoked and the urinary mutagenicity.

Ingestion of foods containing mutagens may be expected to result in excretion of those mutagens. We examined the mutagenicity of urines and feces of humans to determine the effect of ingesting fried ground beef. Both urine and feces became mutagenic in this feeding experiment (Hayatsu et al. 1985a, b, 1986b) (Fig. 9). Analysis of the mutagenic fractions by HPLC indicates that the mutagenic compounds in urine and feces are not those detectable in the fried beef, such as MeIQx (see Sect. 4.1). When rats were fed MeIQx, and the urine and feces of these rats were extracted with blue cotton and the mutagens were fractionated by HPLC, three mutagenic metabolites were isolated: 8-hydroxymethyl-IQx, N-acetylated MeIQx, and N^3-demethyl MeIQx (Hayatsu et al. 1987). Bashir et al. (1987) have

Table 6. Comparison of efficiency in methods for extracting mutagens from smoker's urine (Kobayashi and Hayatsu 1984)[a]

Method	His$^+$ revertants/plate					
	Urine sample					
	HK-1	HK-2	HK-3	SH	YF	TK
Blue cotton	190	216	274	480	392	674
XAD-2 resin	30	175	491	386	158	497
Sep-Pak C_{18}	31	ND	64	69	37	60

[a] Each urine sample (300 ml), which was collected in 1 day, was divided into three 100-ml-portions, and the 100 ml was processed by one of the three methods. Assay was done on *S. typhimurium* TA98 with S9-mix. Solvent controls (dimethylsulfoxide only) gave 32 ± 5 revertants/plate. HK was male, age 22, smoking 10-20 cigarettes per day (the three samples were taken on different days); SH, age 58, 20-30 cigarettes per day; YF, age 56, 20-30 cigarettes per day; TK, age 57, 20-30 cigarettes per day; ND, not done.

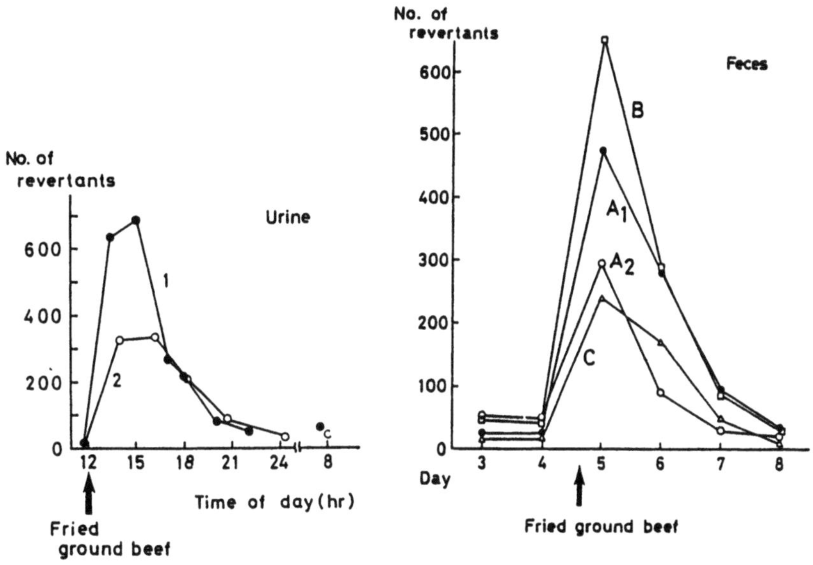

Fig. 9. Mutagenicity of human urine and feces arising from ingestion of fried ground beef. Fried ground beef corresponding to 130–150 g raw meat was eaten, and the blue-cotton extracts of urine and feces were assayed on *S. typhimurium* TA98 with metabolic activation (Hayatsu et al. 1985a, urine; 1985b, feces)

shown that when IQ is treated with human fecal flora and the product is extracted by blue cotton, an oxidized IQ is obtainable; i.e., 2-amino-3,6-dihydro-3-methyl-7*H*-imidazo[4,5-*f*] quinoline-7-one. Blue cotton has also been used in studies on metabolism of IQ in the rat (Inamasu et al. 1988).

In our current monitoring studies of human urine mutagenicity, we have been using blue rayon in place of blue cotton (Hayatsu et al. 1988) because of the higher capacity of blue rayon (see Sect. 2).

4.4 Body Fluids and Tissues

There are reports on this aspect by Manabe and co-workers. These workers detected Trp-P-1 and Trp-P-2 in dialysis fluid of patients with uremia (Manabe et al. 1987a). In this study, a large volume of the dialysis fluid (40 liters) was subjected to the blue cotton extraction and the mutagens in the extract were quantified by HPLC. High recoveries (60–70%) were noted both for Trp-P-1 and Trp-P-2. Likewise, IQ, MeIQ, and MeIQx have been detected in the blue cotton extract of dialysis fluid uremic patients (Yanagisawa et al. 1987). Recent work from this group of researchers has shown that Glu-P-1 and Glu-P-2 can be found in the blue cotton extracts of human plasma (Manabe et al. 1987b). Furthermore, Trp-P-1 and Trp-P-2 can be detected in normal human plasma (Manabe and Wada 1988). An interesting finding is that Glu-P-1 and Glu-P-2 are present in human cataractous lens (Manabe et al. 1988b). The sources of these heterocyclic amines in the human body are not known.

4.5 Water and Air

Monitoring of mutagenicity in the ambient water and air may be facilitated by use of the blue cotton method.

Blue cotton is hung in river water and the mutagenicity is assayed for the methanol-ammonia eluent of the cotton collected after 1 day: in this way, the water of the river Asahi of Okayama was examined, and the mutagenicity, as assayed on *S. typhimurium* TA98 with metabolic activation, was about 2000 revertants per gram blue cotton near the factory area of the city, and 40 revertants per gram at an upstream spot (Hayatsu et al. 1983a).

Since this measurement was made, we have continued this type of work using several different sites of the river to establish a standard procedure to monitor mutagenicity of ambient waters. The following is the one we use currently (Sakamoto and Hayatsu 1988): (1) two or more batches of 1-g blue rayon or blue cotton are placed in meshed nylon bags, together with some weight (usually, pieces of stones); (2) the bags are hung in the water for 24 h; (3) the blue rayon (cotton) is washed with distilled water and eluted with methanol-ammonia (50:1), and then, after drying, is weighed to confirm that there has been no loss of rayon (cotton) during the hanging in water; (4) the mutagenicity is assayed with *S. typhimurium* TA98 and TA100, with and without metabolic activation. In the mutagenicity assay, a dose response (with respect to blue cotton equivalence) is usually measured.

By this method, we have measured the mutagenicity of the rivers Asahi and Sasagase of Okayama and that of the rivers Yodo, Katsura, Kitsu, and Uji of Osaka, and have found that at several spots the activities are very high (up to 4000 revertants in TA98, +S9, per 0.1 g equivalent of blue rayon) (Sakamoto and Hayatsu, unpublished work).

This simple monitoring would provide information as to where and when mutagenicity is present in a given water. With the presently practiced analysis of river water, in which a large volume of water is taken out of the river and processed, the time and labor required to obtain data are enormous.

As I have already discussed (Sect. 2.2), blue cotton is decomposed by the action of chlorine, and therefore is not suitable to exposure to large amounts of tap water which contains chlorine. A trial of using blue cotton to examine mutagenicity of tap water in Finland resulted in detection of no positive activity (Vartiainen et al. 1987).

Blue cotton is also useful in detecting mutagenicity of ambient air. For measuring mutagenicity of airborne particulates, a large volume of air is filtered through a glass fiber, and the particulates on the filter are extracted by sonication in organic solvents. Working on such samples collected in the city of Okayama, we have encountered many cases where linear dose responses were not obtained in the Ames assay; at higher doses extensive cell killings took place. By submitting the samples, suspended in water, to blue cotton extraction, we were able to remove the cytotoxic substances, and linear dose responses were obtained (Iwado et al. 1988).

These studies on the use of blue cotton and blue rayon for the mutagens in water and air are still immature, and much work has to be done for their development.

4.6 Others

As described in the preceding section (Sect. 4.5), a lipophilic sample, such as the air particulate extract, may be subjected to the blue cotton treatment by first preparing an aqueous suspension of the oily material. By using such a manipulation, the mutagenicity of used machine oils has been monitored, and several samples showed positive responses (Kira et al. 1987). Alternatively, such oily materials may be extracted first with aqueous media, and then the extracts can be submitted to the blue cotton treatment.

5 Other Applications

Ethidium bromide is widely used in biochemical laboratories as an agent for visualizing nucleic acids. Since ethidium bromide is a powerful mutagen (Mac-

Gregor and Johnson 1977), its removal from wastes before disposal is desirable. For this purpose, use of blue cotton has been tested. Although blue cotton did adsorb ethidium from its aqueous solutions, the swelling of cotton with the solution precluded the practical use of the cotton for this particular purpose (Lunn and Sansone 1987).

An interesting application of the blue cotton method has been proposed by Povey et al. (1987a, b). They have demonstrated binding of benzo(a)pyrene metabolites in the rat intestinal rumen to magnetic polyethyleneimine microcapsules, and are now investigating the use of copper phthalocyanine tetrasulfonate as an ingredient of the capsules, aiming at making the capsules as a selective, planar-mutagen trap (Povey et al. 1987a).

6 Conclusion and Perspectives

Blue cotton is best suited for use in reducing the masses of samples to be subjected to mutagenicity assays. The river water sample (Sect. 4.5) is an extreme example in this sense. A simple calculation shows that blue cotton (1 g), having a cross-section of 10 cm^2 and a thickness of 1 cm, hung in river water flowing at a rate of 1 cm, would make a contact with 10 ml of water per second, and therefore would make such contacts with a total or 864 liters of water in a period of 24 h.

The power of blue cotton resides in the ligand molecule copper phthalocyanine sulfonate, which is now revealed to have a strong affinity to polycyclic aromatics (Sect. 3.2). However, the chemistry of this affinity has not yet been adequately studied. A systematic search for more powerful ligands than this one seems important in view of the abundance in the environment of mutagens/carcinogens with polycyclic structures.

The carrier of copper phthalocyanine sulfonates may also be changed from cotton (or rayon) to something else. We have prepared blue-cellulose powder and blue-rayon powder, and have used them in chromatographic separation of Trp-P-2 and its metabolites (Kobayashi et al. 1985, and unpublished work). Copper phthalocyanine sulfonates can be bound to anion exchange resins by ionic interactions (Saito et al. 1986). Also, it can be bound through covalent bonds to silica gels (Kobayashi and Hayatsu, unpublished work). These materials may offer their own particular ways of application. We are currently engaged in investigating these possibilities.

Blue cotton is a product of serendipity. Can we hope for some more luck of this kind?

Acknowledgments. I express my sincere gratitude to Dr. T. Sugimura, President of the National Cancer Center, Tokyo, for his continuous encouragement to our study on the use of blue cotton. I also thank Dr. M. Nagao of the National Cancer Center Research Institute for generous gifts of the standard heterocyclic amine samples, with which this work has been promoted greatly.

I am most grateful to my colleagues for their enthusiastic devotion to this study.

This work was supported by Grants-in-Aid (63614523, 62614525, and others) from the Ministry of Education, Science, and Culture, and those (61-shi and others) from the Ministry of Health. Support from Nissan Science Foundation is also acknowledged.

References

Aeschbacher HU, Turesky RJ, Wolleb U, Würzner HP, Tannenbaum SR (1987) Comparison of the mutagenic activity of various brands of food grade beef extracts. Cancer Lett 38: 87–93

Ames BN (1983) Dietary carcinogens and anticarcinogens. Oxygen radicals and degenerative diseases. Science 221: 1256–1264

Ames BN, Durston WE, Yamasaki E, Lee FD (1973) Carcinogens are mutagens: a simple test system combining liver homogenates for activation and bacteria for detection. Proc Natl Acad Sci USA 70: 2281–2285

Ames BN, McCann J, Yamasaki E (1975) Methods for detecting carcinogens and mutagens with the *Salmonella*/mammalian-microsome mutagenicity test. Mutat Res 31: 347–364

Ames BN, Magaw R, Gold LS (1987) Ranking possible carcinogenic hazards. Science 236: 271–280

Arimoto S, Ohara Y, Namba T, Negishi T, Hayatsu H (1980a) Inhibition of the mutagenicity of amino acid pyrolysis products by hemin and other biological pyrrole pigments. Biochem Biophys Res Commun 92: 662–558

Arimoto S, Negishi T, Hayatsu H (1980b) Inhibitory effect of hemin on the mutagenic activities of carcinogens. Cancer Lett 11: 29–33

Bashir M, Kingston DGL, Carman RJ, Van Tassell RL, Wilkins TD (1987) Anaerobic metabolism of 2-amino-3-methyl-3*H*-imidazo[4,5-*f*]quinoline (IQ) by human fecal flora. Mutat Res 190: 187–190

Becher G, Bjørseth A (1983) Determination of exposure to polycyclic aromatic hydrocarbons by analysis of human urine. Cancer Lett 17: 301–311

Blagrove RJ, Gruen LC (1972) The aggregation of the tetrasodium salt of copper phthalocyanine-4,4′,4″,4‴-tetrasulphonic acid. Aust J Chem 25: 2553–2558

Blagrove RJ, Gruen LC (1973) Thermodynamics of the dimerization of copper (II) phthalocyanine-4,4′,4″,4‴-tetrasulphonic acid. Aust J Chem 26: 225–228

Dean CRS, Mather RR, Sing KSW (1978) Gravimetric studies of the sorption of toluene and propanol by copper phthalocyanines. Thermochimica Acta 24: 399–406

Doll R (1977) Strategy for detection of cancer hazards to man. Nature (Lond) 265: 589–596

Edmonds CG, Sethi SK, Yamaizumi Z, Kasai H, Nishimura S, McCloskey JA (1986) Analysis of mutagens from cooked foods by directly combined liquid chromatography-mass spectrometry. Environ Health Perspect 67: 35–40

Felton JS, Knize MG, Shen NH, Lewis PR, Andresen BD, Happe J, Hatch FT (1986) The isolation and identification of a new mutagen from fried ground beef: 2-amino-1-methyl-6-phenylimidazo[4,5-*b*]pyridine (PhIP). Carcinogenesis (Lond) 7: 1081–1086

Friesen M, O'Neill ID, Malaveille C, Garren L, Hautefeuille A, Bartsch H (1987) Substituted hydroxyphenanthrenes in opium pyrolysates implicated in oesophageal cancer in Iran: structures and in vitro metabolic activation of a novel class of mutagens. Carcinogenesis (Lond) 8: 1423–1432

Handbook for Chemical Fibers (1985) JPN Chem Fibers Assoc, p 300

Hayatsu H, Arimoto S, Togawa K, Makita M (1981) Inhibitory effect of the ether extract of human feces on activities of mutagens: inhibition by oleic and linoleic acids. Mutat Res 81: 287–293

Hayatsu H, Oka T, Wakata A, Ohara Y, Hayatsu T, Kobayashi H, Arimoto S (1983a) Adsorption of mutagens to cotton bearing covalently bound trisulfo-copper-phthalocyanine. Mutat Res 119: 233–238

Hayatsu H, Ohara Y, Hayatsu T, Togawa K (1983b) Mutagenicity of chalks. A case in which the test results led to the improvement of the quality of commercial goods. Mutat Res 124: 1–7

Hayatsu H, Matsui Y, Ohara Y, Oka T, Hayatsu T (1983c) Characterization of mutagenic fractions in beef extract and in cooked ground beef. Use of blue cotton for efficient extraction. Gann 74: 472–482

Hayatsu H, Hamasaki K, Togawa K, Arimoto S, Negishi T (1983d) Antimutagenic activity in extracts of human feces. In: Stich HF (ed) Carcinogens and mutagens in the environment, vol II Naturally occurring compounds: endogenous formation and modulation. CRC Press, Boca Raton, Florida, pp 91–99

Hayatsu H, Hayatsu T, Ohara Y (1985a) Mutagenicity of human urine caused by ingestion of fried ground beef. JPN J Cancer Res (Gann) 76: 445–448

Hayatsu H, Hayatsu T, Wataya Y, Mower HF (1985b) Fecal mutagenicity arising from ingestion of fried ground beef in the human. Mutat Res 143: 207–211

Hayatsu H, Kobayashi H, Michi-ue A, Arimoto S (1986a) Affinity of aromatic compounds having three fused rings to copper phthalocyanine trisulfonate. Chem Pharm Bull (Tokyo) 34: 944–947

Hayatsu H, Hayatsu T, Wataya Y (1986b) Use of blue cotton for detection of mutagenicity in human feces extreted after ingestion of cooked meat. Environ Health Perspect 67: 32–34

Hayatsu H, Kasai H, Yokoyama S, Miyazawa T, Yamaizumi Z, Sato S, Nishimura S, Arimoto S, Hayatsu T, Ohara Y (1987) Mutagenic metabolites in urine and feces of rats fed with 2-amino-3,8-dimethylimidazo[4,5-*f*]quinoxaline, a carcinogenic mutagen present in cooked meat. Cancer Res 47: 791–794

Hayatsu H, Hayatsu T, Zheng QL, Ohara Y, Arimoto S (1988) Problems in monitoring mutagenicity of human urine. In: Bartsch H, Hemminki K, O'Neill IK (eds) Methods for detecting DNA damaging agents in humans: applications in cancer epidemiology and prevention. Lyon pp 401–404 (IARC Sci Publ No 89)

Inamasu T, Luks HJ, Weisburger JH (1988) Comparison of XAD-2 column and blue cotton batch techniques for isolation of metabolites of 2-amino-3-methylimidazo[4,5-*f*]quinoline. JPN J Cancer Res (Gann) 79: 42–48

Iwado H, Naito M, Hayatsu H (1988) Mutagenicity of amibient air particulates detected by the blue cotton method. Abstr 108th Annu Meet Pharm Soc JPN, p 646

Jägerstad M, Olsson K, Grivas S, Negishi C, Wakabayashi K, Tsuda M, Sato S, Sugimura T (1984) Formation of 2-amino-3,8-dimethylimidazo[4,5-*f*]quinoxaline in a model system by heating creatinene, glycine and glucose. Mutat Res 126: 239–244

Kasai H, Yamaizumi Z, Nishimura S, Wakabayashi K, Nagao M, Sugimura T, Spingarn NE, Weisburger JH, Yokoyama S, Miyazawa T (1981a) A potent mutagen in broiled fish. Part 1. 2-Amino-3-methyl-3*H*-imidazo[4,5-*f*]quinoline. J Chem Soc Perkin Trans 1: 2290–2293

Kasai H, Yamaizumi Z, Shiomi T, Yokoyama S, Miyazawa T, Wakabayashi K, Nagao M, Sugimura T, Nishimura S (1981b) Structure of a potent mutagen isolated from fried beef. Chem Lett: 485–488

Kato T, Kikugawa K, Hayatsu H (1986) Occurrence of the mutagens 2-amino-3,8-dimethylimidazo[4,5-*f*]quinoxaline (MeIQx) and 2-amino-3,4,8-trimethylimidazo[4,5-*f*]quinoxaline(4,8-Me_2IQx) in some Japanese smoked, dried fish products. J Agric Food Chem 34: 810–814

Kikugawa K, Kato T (1987) Formation of mutagens, 2-amino-3,8-dimethylimidazo[4,5-*f*]quinoxaline (MeIQx) and 2-amino-3,4,8-trimethylimidazo[4,5-*f*]quinoxaline (4,8-Di-MeIQx), in heated fish meats. Mutat Res 179: 5–14

Kikugawa K, Kato T, Hayatsu H (1985) Mutagenicity of smoked, dried bonito products. Mutat Res 158: 35–44

Kikugawa K, Kato T, Hayatsu H (1986a) The presence of 2-amino-3,8-dimethylimidazo-[4,5-*f*]quinoxaline in smoked dry bonito (*katsuobushi*). JPN J Cancer Res (Gann) 77: 99–102

Kikugawa K, Kato T, Hayatsu H (1986b) Formation of mutagenic substances during smoking-and-drying (*baikan*) of bonito meat. Eisei Kagaku 32: 379–383

Kikugawa K, Kato T, Takahashi S (1989) Possible presence of 2-amino-3,4-dimethylimidazo [4,5-*f*]quinoline (MeIQ) and other heterocyclic amine-like mutagens in roasted coffee beans. J Agric Food Chem 37: 881–886

Kira S, Hayatsu H, Nogami Y, Ogata M (1987) Mutagenicity in machine oils. JPN J Ind Health 29: 296–297

Kobayashi H, Hayatsu H (1984) A time-course study on the mutagenicity of smoker's urine. Gann 75: 489–493

Kobayashi H, Ishizu Y, Hayatsu H (1985) Use of blue cellulose in fractionation of mutagens. Mutat Res 147: 261

Lunn G, Sansone EB (1987) Ethidium bromide: destruction and decontamination of solution. Anal Biochem 162: 453–458

MacGregor JT, Johnson IJ (1977) In vitro metabolic activation of ethidium bromide and other phenanthridinium compounds: mutagenic activity in *Salmonella typhimurium*. Mutat Res 48: 103–107

Manabe S, Wada O (1988) Analysis of human plasma as an exposure level monitor for carcinogenic tryptophan pyrolysis products. Mutat Res 209: 33–38

Manabe S, Yanagisawa H, Guo SB, Abe S, Ishikawa S, Wada O (1987a) Detection of Trp-P-1 and Trp-P-2, carcinogenic tryptophan pyrolysis products, in dialysis fluid of patients with uremia. Mutat Res 179: 33–40

Manabe S, Yanagisawa H, Ishikawa S, Kitagawa Y, Kanai Y, Wada O (1987b) Accumulation of 2-amino-6-methyldipyrido[1,2-*a*:3′,2′-*d*]imidazole and 2-aminodipyrido[1,2-*a*:3′,2′-*d*]imidazole, carcinogenic glutamic acid pyrolysis products, in plasma of patients with uremia. Cancer Res 47: 6150–6155

Manabe S, Kanai Y, Yanagisawa H, Tohyama K, Ishikawa S, Kitagawa Y, Wada O (1988a) Detection of carcinogenic glutamic acid pyrolysis products in Worcestershire sauce by high-performance liquid chromatography. Environ Mol Mutagen 11: 379–388

Manabe S, Yanagisawa H, Kanai Y, Wada O (1988b) Presence of carcinogenic glutamic-acid pyrolysis products in human cataractous lens. Ophthalmic Res 20: 20–26

Mohtashamipur E, Norpoth K, Lieder F (1985) Isolation of frameshift mutagens from smoker's urine: experiences with three concentration methods. Carcinogenesis (Lond) 6: 783–788

Moser FH, Thomas AL (1983) The Phthalocyanines, vol I Properties, vol II Manufacture and Applications. CRC Press, Boca Raton, Florida

Muramatsu M, Matsushima T (1985) Formations fo MeIQx and 4,8-DiMeIQx by heating mixtures of creatinine, amino acids and monosaccharides. Mutat Res 147: 266

Muraoka N, Yasuda K, Inagaki H (1986) Mutagenicity of cooked foods, V. Mutagenicity of grilled rice. Mutat Res 164: 275

Nagao M, Wakabayashi K, Sugimura T (1985) Mutagens in food and drinks, and their carcinogenicity. In: Zimmerman and Taylor (eds) Mutagenicity testing in environmental pollution control. Ellis Horwood Limited, Chichester, pp 69–85

Negishi C, Wakabayashi K, Tsuda M, Sato S, Sugimura T, Saito H, Maeda M, Jägerstad M (1984a) Formation of 2-amino-3,7,8-trimethylimidazo[4,5-*f*]quinoxaline, a new mutagen, by heating a mixtures of creatinine, glucose and glycine. Mutat Res 140: 55–59

Negishi C, Wakabayashi K, Tsuda M, Saito H, Maeda M, Sato S, Sugimura T, Jägerstad M, Muramatsu M, Matsushima T (1984b) Formation of MeIQx and a new mutagen DiMeIQx on heating mixtures of creatinine, glucose and amino acids. Environ Mutagen Res Commun 6: 129–136

Negishi C, Wakabayashi K, Yamaizumi Z, Saito H, Sato S, Sugimura T, Jägerstad M (1985) Identification of 4,8-DiMeIQx, a new mutagen. Mutat Res 147: 267

Okuda T, Yoshida T, Mori K, Hatano T (1981) Tannins of medicinal plants and drugs. Heterocycles 15: 1323–1348

Povey AC, Brouet I, Bartsch H, O'Neill IK (1987a) Binding of benzo[a]pyrene metabolites in the rat intestinal lumen by magnetic polyethyleneimine microcapsules following an intragastric dose of [^{14}C]benzo[a]pyrene. Carcinogenesis (Lond) 8: 825–831

Povey AC, Nixon JR, O'Neill IK (1987b) Trapping of chemical carcinogens with magnetic polyethyleneimine microcapsules: II. Effect of membrane and reactant structures. J Pharm Sci 76: 194–200

Saito Y, Mifune M, Kawaguchi T, Odo J, Tanaka Y, Chikuma M, Tanaka H (1986) Catalase-like catalytic activity of ion-exchange resins modified with metalloporphyrins. Chem Pharm Bull (Tokyo) 34: 2885–2889

Sakamoto H, Hayatsu H (1988) Concentration of mutagenic components in river water by use of adsorbents bearing covalently bonded copper phthalocyanine. Abstr 108th Annu Meet Pharm Soc JPN: 667

Scheller KH, Sigel H (1983) Molecular properties of 1,N^6-ethenoadenosine in comparison with adenosine: self-association, protonation, metal ion complexation, and tryptophan-adduct formation. A study on ε-adenosine using proton nuclear magnetic resonance, ultraviolet spectrophotometry, and potentiometric pH titration. J Am Chem Soc 105: 3005–3014

Schröder J (1982) Energetische Charakterisierung organischer Pigmente durch Adsorptionsisothermen. Farbe Lack 88: 100–105

Shioya M, Wakabayashi K, Yamashita K, Nagao M, Sugimura T (1989) Formation of 8-hydroxydeoxyguanosine in DNA treated with fecapentaene-12 and -14. Mutat Res 225: 91–94

Sugimura T (1982) Carcinogens. Chuko Shinsho (Chuko New Books) vol 670, Chuo Koron Publishing, Co, Tokyo

Sugimura T (1985) Carcinogenicity of mutagenic heterocyclic amines formed during the cooking process. Mutat Res 150: 33–41

Sugimura T (1988) Successful use of short-term test for academic purposes: their use in identification of new environmental carcinogens with possible risk for humans. Mutat Res 205: 33–39

Tekahashi M, Wakabayashi K, Nagao M, Yamamoto M, Masui T, Goto T, Kinae N, Tomita I, Sugimura T (1985a) Quantification of 2-amino-3-methylimidazo[4,5-*f*]quinoline (IQ) and 2-amino-3-8-dimethylimidazo[4,5-*f*]quinoxaline (MeIQx) in beef extracts by liquid chromatography with electrochemical detection (LCEC). Carcinogenesis (Lond) 6: 1195–1199

Takahashi M, Wakabayashi K, Nagao M, Yamaizumi Z, Sato S, Kinae N, Tomita I, Sugimura T (1985b) Identification and quantification of 2-amino-3,4,8-trimethylimidazo[4,5-*f*] quinoxaline (4,8-DiMeIQx) in beef extract. Carcinogenesis (Lond) 6: 1537–1539

Takahashi M, Wakabayashi K, Nagao M, Tomita I, Sugimura T (1985c) Quantification of mutagenic/carcinogenic heterocyclic amines in cooked foods. Mutat Res 147: 275

Turesky RJ, Bur H, Huynh-Ba T, Aeschbacher HU, Milon H (1988) Analysis of mutagenic heterocyclic amines in cooked beef products by high-performance liquid chromatography in combination with mass spectrometry. Fd Chem Toxic 26: 501–509

Vartiainen T, Lematainen A, Jääskeläinen S, Kauranen P (1987) Comparison of solvent extractions and resin adsorption for isolation of mutagenic compounds from chlorinated drinking water with high humus content. Water Res 21: 773–779

Venitt S (1988) The use of short-term test for the detection of genotoxic activity in body fluids and excreta. Mutat Res 205: 331–353

Yamasaki E, Ames BN (1977) Concentration of mutagens from urine by adsorption with the nonpolar resin XAD-2: cigarette smokers have mutagenic urine. Proc Natl Acad Sci USA 74: 3555–3559

Yamashita M, Wakabayashi K, Nagao M, Sato S, Yamaizumi Z, Takahashi M, Kinae N, Tomita I, Sugimura T (1986a) Detection of 2-amino-3-methylimidazo[4,5-*f*]quinoline in cigarette smoke condensate. JPN J Cancer Res (Gann) 77: 419–422

Yamashita M, Wakabayashi K, Nagao M, Sato S, Kinae N, Tomita I, Sugimura T (1986b) Amounts of heterocyclic amines in basic fraction of cigarette smoke condensates. Mutat Res 164: 286

Yanagisawa H, Manabe S, Wada O (1987) Detection of IQ-type heterocyclic amines in dialysis fluid of uremic patients treated by peritoneal dialysis. JPN J Nephrology 29: 61–67

Zhang XM, Wakabayashi K, Liu ZC, Sugimura T, Nagao M (1988) Mutagenic and carcinogenic heterocyclic amines in Chinese cooked food. Mutat Res 201: 181–188

Meiosis in Mammals and Man

R. M. SPEED[1]

Contents

1 Introduction

Meiosis in mammals and man, the process whereby haploid male and female germ cells are produced by a specialized reduction division, is still a relatively recent discovery. Little over 100 years have elapsed since Mendel (1865) proposed that the determinants of phenotype were recombined and transmitted by the meiotic chromosomes. The origins of meiosis and recombination, and their place in the evolution of mammalian sexual reproduction, are still to be clarified. Maynard-Smith (1978) has examined the possibility that meiotic recombination is purely a mechanism for the repair of mutational damage. Against this, one must remember that in many species of male *Diptera,* recombination has been eliminated. The advantages of mutational repair within these systems must then have been small, in comparison to those factors removing the process.

For the next few decades after Mendel mammalian cytogenetics concentrated on mitotic chromosomes and the problems of determining the diploid chromosome number in a variety of animal species, mainly due to poor preparative techniques. It was not until Ford and Hamerton (1956) used an improved squash technique that the meiotic complement of the human spermatocyte was shown to consist of

[1] MRC Human Genetics Unit Western General Hospital, Crewe Road, Edinburgh, Great Britain EH4 2XU

23 bivalents. Subsequently, very little experimental meiotic data has been available from humans because of the obvious ethical problems.

The majority of experimental studies on meiotic chromosomes have occurred in mammals, and in particular mice and rats. Such diverse subjects as fertility have been reviewed by Searle (1982) and De Boer (1986); mutation, either spontaneous or induced by X-rays or chemicals, by Kimball (1987); and aneuploidy by Bond and Chandley (1983), to name but a few. A limited amount of experimental meiotic investigations have also been undertaken in primates. Breeding studies in lemur have allowed an analysis of the spermatogenic system in interspecific hybrids (Ratomponirina et al. 1982), while the effects of X-ray damage on male marmoset germ cells have been studied by Brewen et al. (1975) and female chiasma frequencies have been analyzed in species of *Macaca* and *Nemestrina* (Jagiello et al. 1973).

Data on human meiotic chromosome structure, synapsis, and disjunction has nonetheless been acquired from testicular biopsies, obtained from male patients attending subfertility clinics (Skakkebaek et al. 1973; Koulischer and Schoysman 1975; Chandley et al. 1976), and ovaries from aborted female foetuses (Luciani and Stahl 1971; Wallace and Hulten 1985; Speed 1988). These materials have given an initial, but limited, insight into the problems of human meiosis at both light and electron microscope levels.

1.1 Differences Between Male and Female Meiosis

While being processes of similar function, that is, the production of haploid gametes, it is important to realize that male and female mammalian meiosis are of a very different temporal nature. In general, germ cells in both mammalian sexes migrate from regions near the yolk sac endoderm to areas overlying the mesonephros which are destined to become the foetal gonads. Differences between the sexes then become apparent.

Only mitotic activity is observed within the embryonic testis and after a few cell cycles the germ cells invade the tubules and become resting gonocytes. Further progression is delayed until puberty, when gonocytes progressively become type A and B spermatogonia, 1° and 2° spermatocytes, and finally, spermatids. The complex, overall genetic control of spermatogenesis and the question of haploid gene expression, of which in reality little is as yet known, has been comprehensively reviewed by Handel (1987).

By contrast activity in the foetal ovary consists of the oogonia continuing development and greatly increasing in number by mitotic division. The numbers of germ cells present at mid-term varies considerably between species. In humans, 7 million germ cells can be compared with 3.5 million in monkeys, and 150 000 in rats. By term, the numbers of germ cells have markedly declined by degeneration (atresia), only 2 million remaining in the human (Baker 1963). When we consider that only 400 to 500 human oocytes will finally be ovulated in adult life, the initial gross overproduction of oocytes and their drastic selection by atresia only serves to contrast the male and female system. After the final oogonial mitotic interphase, premeiotic DNA synthesis occurs, and the prophase of meiosis can be observed as early as week 11 of gestation in the human female foetus (Speed 1985). By the time of birth mammalian oocytes generally enter the so-called resting dictyotene stage. However, just after birth when the oocyte and its enclosing follicle grow in diameter, active RNA transcription by the lampbrush-like chromosomes occurs, such RNA possibly being laid down for later stages of meiosis. The dictyate nucleus can only truly be said to be in the resting stage when the granulous cells of the follicle begin to supply maternal proteins to the germ cell. This stage then persists until puberty in the female, and in its extreme form may last for 12 years in humans. Follicular maturation then takes place, and a certain percentage of oocytes are recruited to respond to pituitary gonadotrophins at each ovulatory cycle. In response to follicle stimulating hormone, some undergo final maturation, and progress through diplotene and metaphase I by the action of leutenizing hormone. The first meiotic division is completed with the production of the metaphase II chromosomes and the first polar body. Only on fertilization does further development occur when the second meiotic division is initiated, resulting in chromatid separation and second polar body formation.

Several important features therefore distinguish male and female mammalian meiosis: (1) while spermatogenesis is an ongoing process with continued multiplication from basic stem cells, the number of oocytes in female mammals is fixed at birth; (2) prophase synapsis and recombination will have taken place in the female before birth; (3) while male meiosis is completed in a relatively short time, varying from 12 days in mouse to approximately 24 days in man (Monesi 1972), in the female it is spread out over a much longer period. In the female mouse, meiosis can vary in duration from 2 to 12 months, while in human the period may be from ten to forty years (Baker 1972). Lastly, considering problems of male and female fertility, the extrusion of two degenerative polar bodies during female meiosis allows the elimination of abnormal chromosome complements. This is not possible in the male where the four products of meiosis are initially viable sperm.

1.2 Mechanism of Synapsis

Meiotic chromosome pairing in most mammalian species commences with the aligment of the lateral elements (roughly 300 nm apart) of homologous chromosomes. When this distance is reduced to approximately 100 nm a synaptonemal complex (SC) begins to form. Initiation points may be numerous, as in plant species, or virtually all telomeric, as in the human oocyte. Normally only homologous synapsis between identical chromosome pairs occurs at zygotene, but numerous examples now exist of SCs that originate from chromosomes or segments of chromosomes that are non-homologous (heterologous pairing) in genetic content (Von Wettstein et al. 1984; [Speed 1986b]). Heterologous pairing may also develop in chromosomal rearrangements such as duplications or inversions at prophase and has been termed synaptic adjustment by Moses et al. (1978). As such SCs appear of normal dimensions and structure it would appear that the SC per se is not a structure that is responsible for recombination, as such heterologous pairing in plant haploid species (Sadasiviah and Kasha 1971) does not lead to chiasma formation or crossing-over.

It is generally held that recombination will only take place when DNA sequences of strict homology are brought into register either in the central region of the SC, or within the bulk of the chromatin surrounding the SC. More than 99% of the chromosomal DNA remains outside the confines of the SC, it being calculated for *Neurospora* (Westergaard and Von Wettstein 1972) that the total nuclear SC length of 50 μ represents only 0.3% of the total DNA double helix length of 16 mm. Gillies (1973) has shown that the corresponding figure for maize is only 0.014% of the total DNA length. Experimental evidence, however, has favoured the proposal that recombination occurs within the SC, primarily because Westergaard and Von Wettstein (1970) have observed that chiasma appear to cross remnants of the diplotene SC in the fungus *Neottiella.* Secondly, Moses and Poorman (1984) have found an apparent association between P-DNA synthesis and the central regions of the SC in mouse pachytene oocytes. Such observations are thought to represent evidence of a breakage and reunion mechanism associated with crossing-over during the pachytene stage of meiosis. Alternatively, it has been suggested by Riley and Flavell (1977) that meiotic homologue synapsis may only initially involve the matching of isolated chromatin blocks or DNA sequences scattered along the chromosomes. In this way, chromosomes which are only similar in a gross manner could at least synapse, if not be capable of recombination.

The various mechanisms proposed to explain synapsis have recently been reviewed by Chandley (1986), who has further suggested that early replicating R-band sites could act as pairing initiation points for both homologous and non-homologous meiotic synapsis. Heterologous SC synapsis may, however, serve other functions, perhaps related to fertility, rather than to those ultimately leading to recombination in germ cells. In proposing a theory relating chromosome pairing and fertility in species as varied as *Drosophila* and the human, Miklos (1974) had noted a correlation between the reduction or absence of synapsis and gametogenic breakdown. This led to his proposal that saturation of pairing sites between homologues of either sex chromosomes or autosomes was vital for the normal progression of germ cell development. Burgoyne and Baker (1984) later extended this model to include both mammalian male and female germ cells. Many such examples of the operation of this theory have been reported, and Burgoyne and Biddle (1980) suggested that the loss of spermatocytes in XYY mice and human males was due to the univalence of the X and Y chromosomes.This effect has also recently been observed in a male patient carrying a deletion in the vicinity of the pseudoautosomal region of the X short arm at p22.32 involving the loss of the Xg, M1C2X and the STS loci (Curry et al. 1984). The X and Y chromosome in this patient fail to pair at prophase, remaining as univalents in 100% of cells, and spermatogenesis then fails at the MI stage (Speed RM, 1988, unpublished observation). Also, human males with either a ring-21 (McIlree et al. 1966) or ring-E group chromosome (Kjessler 1974) show a strong correlation between the failure to pair of a single autosomal bivalent and male sterility. A question as to whether the fertility of an organism might be influenced if heterologous synapsis, as opposed to homologous synapsis, was permissible in chromosomally unbalanced situations can be answered by comparing human and mouse XO females. In human XO situations, the single X fails to pair in any way and sterility is the norm for adult women. In XO mice the X chromosome either pairs heterologously with itself or with other autosomes, and fertility, while reduced in comparison with control XX female mice, is usual.

2 Cytological Technology from Squash to Surface Spreading

Two major advances have occured since Ford and Hamerton (1956) employed a squash technique to establish that human spermatocytes contained 23 bivalents. The development of a quick reliable air-drying method, utilizing a hypotonic shock and methanol/acetic acid fixation by Evans et al. (1964) allowed a detailed analysis of the meiotic divisions in a great variety of mammalian species. More recently, the modification of the Counce and Meyer (1973) surface-spreading technique to mammalian species has permitted a detailed analysis with both light (Fletcher 1979) and electron microscopy. (Moses 1977a) of meiotic pairing at the synaptonemal complex (SC) level during prophase.

2.1 Air Drying

The stages of meiosis from prophase to metaphase I and II have been examined in various mammalian species (Fig. 1). and clearly defined in man by Chandley (1975). Modification of somatic chromosome staining techniques have allowed the location of meiotic centromeres (Fig. 1d-f) Chandley and Fletcher 1973) and chiasma position (Hultén 1974) at metaphase I. The isolation of the X and Y chromosomes within the sex vesicle at pachytene has been investigated in the mouse by Solari (1970) and in the human by Solari and Tres (1970). The sex bivalent has been shown to be condensed, late replicating and undergoing an early cessation of transcription (Fig. 1a) relative to the autosomes (Tres 1975). Chromomere maps have been constructed following specific staining techniques at the pachytene stage (Hungerford 1971; Ambros and Sumner 1987). The banded appearance of the meiotic bivalents show close approximation to the G-banding patterns seen in somatic chromosomes of mitotically dividing cells.

In the female mammal, the separation of the initial pachytene stages in the foetal ovary from the later meiotic divisions in the adult has necessarily led to modifications of the techniques used in the study of male meiosis. Simple squash preparations initially allowed the identification of the prophase stages (Ohno et al. 1961). Chromomere patterns and further clarity of synapsis (Fig. 1b) was made possible by the air-drying methods of Luciani and Stahl (1971). The study of first (Fig. 1e) and second metaphase chromosomes in oocytes was only possible when methods of in vitro culture of mature oocytes were developed. Data on the effectiveness of pairing and disjunction and chiasma frequency have been obtained

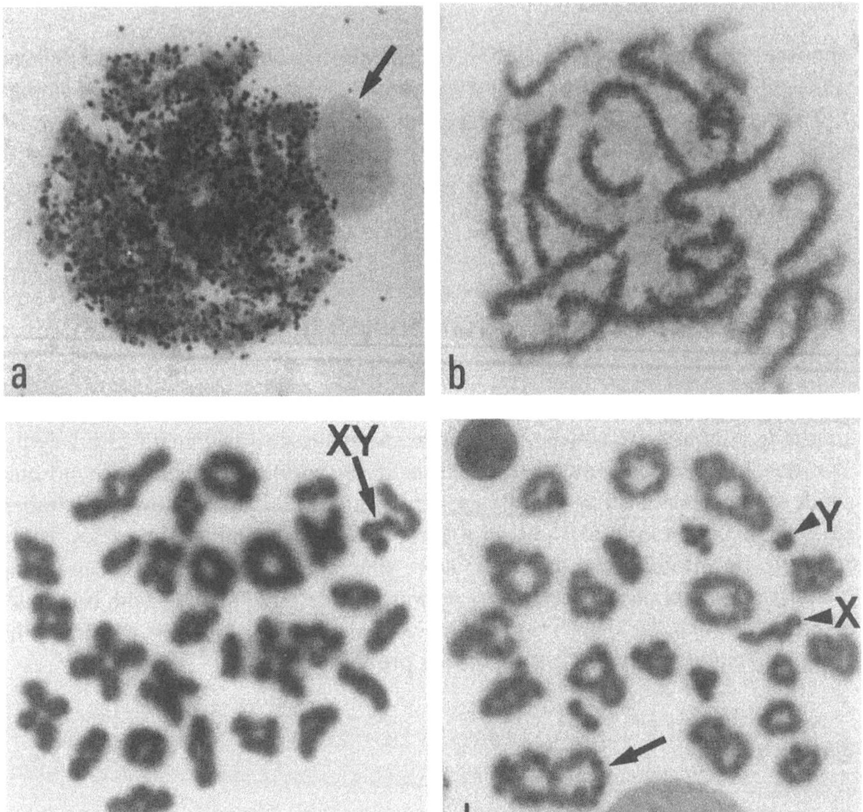

Fig. 1 a–f. Male and female air-dried meiotic preparations. **a** Mouse pachytene, male, ^{3}H uridine labelled. Active autosomal RNA transcription, but sex vesicle *(arrow)* unlabelled; **b** mouse pachytene, female, after extended hypotonic treatment showing individual bivalents; **c** bank vole *(Clethrionomys glareolus)* MI, male. Conventional Giemsa staining; 2n–56. **d** Human MI, male. C-banded preparation. X and Y detached *(arrowheads),* translocation quadrivalent *(arrow);* 2n–46. **e** Mouse MI, female. C-banded preparation; 2n–40. **f** Mouse MII, male. C-banded preparation

for the female mouse (Henderson and Edwards 1968) and monkey (Jagiello et al. 1973). Extending the technique to the human has proved difficult: firstly, because of the ethical difficulties in obtaining mature human oocytes; secondly, because human oocytes fail to provide analyzable bivalents in culture. Indeed, throughout the scientific literature very few reports of clear human metaphase I or II chromosomes exist (Jagiello et al. 1968; Yuncken 1968; Chandley 1971; Uebele-Kallhardt and Knörr 1971; Polani et al. 1982). For the human female, it seems that the chiasma frequency is lower in oocytes than in male spermatocytes (Chandley 1975), whereas in other mammalian species, where more extensive data are available, the reverse would appear to be true (Polani 1972).

2.2 Surface Spreading

The introduction of the surface-spreading technique to mammalian meiosis has led to a rapid expansion of data on synapsis in many mammalian species. Information from the male has predominated because of the more readily available testicular tissue, whether from experimental animals or from human male patients attending subfertility clinics. Studies at our laboratory at Edinburgh have included human infertility and aneuploidy as major subjects of interest. The results of the new surface-spreading technique, as applied to both human male and female meiosis, will constitute a large portion of this chapter.

Initial observations in the male have shown that unless chromosomal mutations are present, autosomal pairing is extremely regular in both mice and man. More attention has been paid to the staging of the male meiotic prophase, utilizing changes in morphology of the lateral elements, the nucleoli, and in particular, the behaviour of the sex chromosome axes. In this way the sequence of male prophase development has been determined for the mouse (Moses 1981), the hamster (Moses 1977b), the rat (Joseph and Chandley 1984), and the human (Solari 1980).

In the human male, six basic substages have been identified, ranging from type O at late zygotene where no apparent SC has formed between the X and Y chromosomes, to type V at late pachytene. An initially limited homologous pairing at stage I (Fig. 2a) is followed by maximum homologous and heterologous synapsis at Stage II (Fig. 2b). During Stage III and IV, the X and Y SC decreases in length and the unpaired axial elements become multiply split. By Stage V, the XY bivalent is represented by a tangled mass with no clearly identifiable pairing region. A more detailed description of the human XY pairing types has recently been given by Chandley et al. (1984), who proposed that pairing in humans could extend across the centromere, and might progress to include the entire euchromatic segment of the Y chromosome as is normally seen in rat (Fig. 2c). This was recently confirmed in the human male by Sumner and Speed (1987), using an immunochemical labelling technique which allowed the kinetochores to be accurately labelled within the XY bivalent, showing that pairing did extend beyond the Y centromere. To what extent such pairing between the human X and Y chromosomes is truly homologous, as opposed to heterologous, is currently a matter for debate, involving questions as to the origins of the heteromorphic sex-

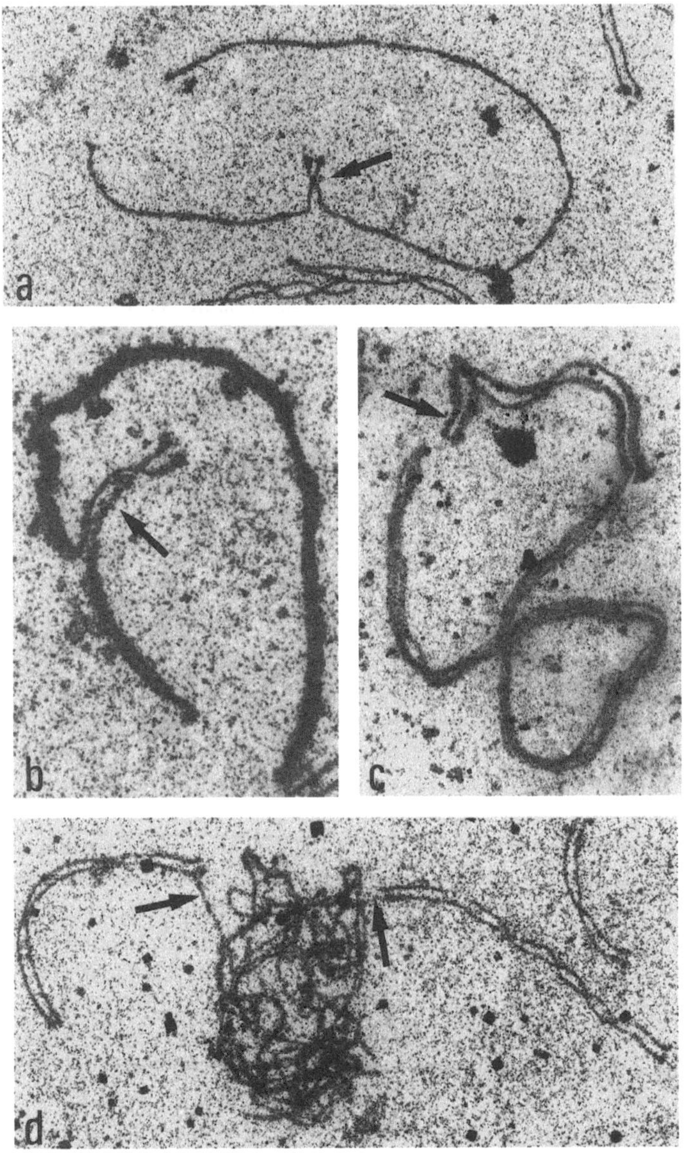

Fig. 2 a–d. XY stages and pairing at the EM level. **a** Type I human XY bivalent. Short SC *(arrow)* which includes the pseudo-autosomal region. **b** Type II human XY bivalent. Half the Y chromosome included in SC *(arrow)*, much of which represents heterologous synapsis. **c** Rat XY bivalent, showing Y chromosomes *(arrow)*, completely paired with X. **d** Type IV human XY bivalent showing contact *(arrows)* with two probable acrocentric autosomes

determining chromosomal systems of mammals, and as to why they are segregated within the sex vesicle and out of synchrony with the autosomal complement.

Koller and Darlington (1934) had initially proposed that at least a segment of homologous pairing occured between the X and Y, within which at least one cross-over could occur. This was further supported by Ohno et al. (1959), who suggested that the end-to-end associations seen in many mammalian XY sex chromosome systems resulted from the terminalization of an already nearly terminal chiasma. Burgoyne (1982) has proposed that pairing between the X and Y chromosome is to a certain point a consequence of homology, albeit confined to a small terminal segment. An obligatory cross-over occurs within such a region, and genes distal to such a chiasma will be inherited in a pseudo-autosomal manner. Confirmation of such events is however difficult, as the mapping of such genes will be extremely complicated on the basis of regular pedigree analysis. An example of an apparently autosomally inherited mutant, for which an autosomal location has been ruled out, is the sex reversing (Sxr) mutation described by Cattanach et al. (1971) in the mouse. The mutation has involved the duplication and transference of the testis-determining factor from a proximal location on the Y chromosome to the distal tip beyond the normal pairing region. The Sxr fragment is then transferred by an exchange event at meiosis to the X chromosome, determining that females inheriting such an X will be phenotyptically male. Evans et al. (1982) have cytologically confirmed that such an exchange does occur. A consistent feature at metaphase I in such Sxr males is the high proportion of cells 70–90% with an unpaired X and Y, compared with 5–10% in controls. A recent study at the EM level (Chandley and Speed 1987) has shown that the Y chromosome with two testis-determining sequences has a strong drive to initiate self-pairing and completely synapse heterologously at the expense of regular synapsis with the X chromosome. In this case, the heterologous pairing of the Y may contribute to the varying fertility in Sxr males, as an association between spermatogenetic failure and lack of X and Y chromosome pairing has long been known (Beechey 1973).

Ashley (1984) has, however, taken the view that, apart from "end attraction", synapsis between the X and Y chromosomes is mostly heterologous in nature, and taken in combination with a premature prophase desynapsis of the sex chromosomes, is the basis of a mechanism that has evolved to maintain regular disjunction, but to prevent crossing-over. Mutants such as the Sxr male mouse have to be explained on the basis of an aberrant delay in desynapsis, which will permit such exchanges to take place between the X and Y.

Recent advances in molecular genetics suggest that a combination of the above proposals is permissible. DNA probes now indicate that a region of homology exists between the mammalian sex chromosomes. The MIC2X and Y DNA sequences both map to the distal ends of the X and Y short arms (Goodfellow et al. 1985) as are the hypervariable telomeric sequences reported by Cooke et al. (1985) and Simmler et al. (1985). As expected by the Burgoyne theory, the latter sequences show pseudo-autosomal inheritance. This region is, however, small in relation to the distance over which the X and Y can pair as visualized by electron microscopy. The testis-determining factor (Tdf) in the human male has been relocated by Goodfellow et al. (1985) from a pericentromeric region to the short arm of the Y chromosome. While this factor is outside the pseudo-autosomal region, occasional XX males whose DNA profiles contain X-specific sequences are detected, presumably because an exchange event as originally proposed by Ferguson-Smith (1966) takes place between the X and Y chromosome. Indeed, a gradient of recombination decreasing from the telomeres of the X and Y has been reported by Rouyer et al. (1986), which might permit such rare cross-overs causing the Tdf exchange. Thus the majority of mammalian XY synapsis appears to be heterologous, preventing wide-scale recombination, but a small telomeric region does exist, where pseudo-autosomal exchange between common loci can and does occur.

Turning to female germ cells, surface spreading as in the male has permitted a detailed analysis of their progression through prophase in the mouse (Moses and Poorman 1984) and man (Speed 1985). Chromosomally normal foetal ovaries can be obtained from social abortions, and such material has allowed a comparison of oocyte degeneration (atresia) with chromosomally abnormal situations as in XO and trisomy 18 and 21 human foetuses (Speed 1986a and b, 1988) where arrest is known to be more severe. It has also been suggested that synaptic irregularities in the foetal ovary are the cause of the high levels of non-disjunction seen in later reproductive life in adult females (Henderson and Edwards 1968). The so-called Production Line theory has also been examined by surface spreading in foetal ovaries (see below).

The female germ cell lacks a characteristic XY bivalent (the XX bivalent behaving exactly as an autosomal bivalent); this has meant that the classification of germ cell progression has relied on the estimation of the length of SCs, synaptic and desynaptic behaviour, and nucleolar morphology (Moses and Poorman 1984).

Studies at the EM level by Speed (1988) in chromosomally normal human foetal oocytes have shown high levels of synaptic disturbance when compared with male spermatocytes (Table 1). Similar results have been reported by Mahadevaiah and Mittwoch (1986) in the mouse. Only 54% of human oocytes studied at Edinburgh showed normal synapsis. Atretic cells (Z cells) occured three times more frequently than in normal males. Categories of abnormality seen in the

Table 1. Normal and abnormal cell types at EM level in chromosomally normal males and females (%)

	Normal males	Subfertile males	Normal females
No. individuals	13	32	5
Total No. cells scored	549	1578	1200
Normal	92.2	78.4	53.6
Z cells	4.7	13.0	15.0
Interlocks	0.2	0.2	1.5
Non-homologous pairing	0.2	0.4	7.6
Triple pairing	0.0	0.0	4.3
Interchange	0.0	0.1	2.2
Asynaptic region	2.0	3.2	7.1
Univalents	0.4	0.6	7.8

female include partial asynapsis of bivalents, which in a more extreme form leads to total asynapsis and univalent formation. This was 20 times more frequent in human oocytes compared with male germ cells. Pairing between non-homologues also oocurs with a high frequency in human female germ cells. In its simplest form, small sections of lateral elements pair with themselves, forming hairpins or loops; at the extreme, virtually complete heterologous synapsis can occur. For example, a pair of SCs, having one normally synapsed set of telomeric ends, shows the telomeres at the other end of the SC in an asymmetrical orientation. Two

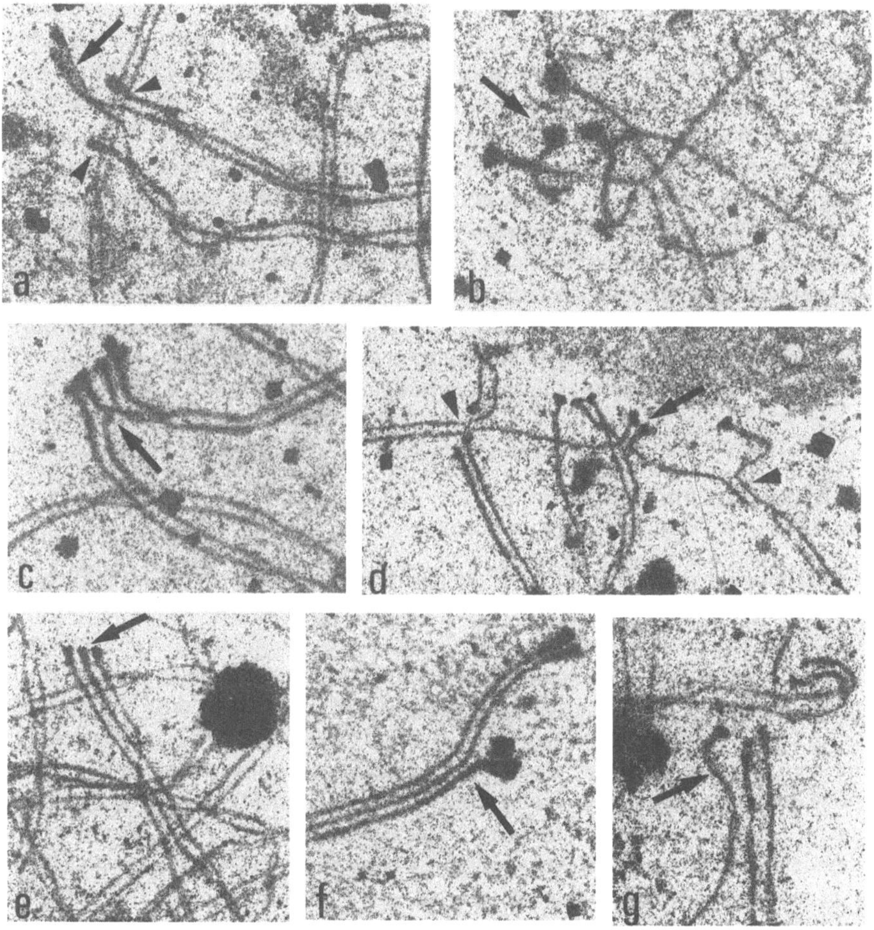

Fig. 3 a–g. Heterologous synapsis in human foetal oocytes. **a** Almost complete heterologous synapsis between pairs of lateral elements of differing lenths *(arrow* cf. *arrowheads)*. Probable terminal homologous pairing between largest elements *(arrow)*. **b** Telomere clustering at early zygotene. **c** Subtelomeric exchange of pairing partner. **d** Possible origin of exchange as in *c*. Homologous pairing from subtelomeric region *(arrowheads)* with heterologous synapsis at telomere *(arrow)*. **e** Triple pairing *(arrow)*, telomeres in register. **f** Triple pairing with extra lateral element *(arrow)* asymmetrically aligned. **g** Possible initiation of triple pairing at early zygotene. With permission of Human Genetics (Speed 1988)

shorter lateral elements have apparently fully paired with two longer lateral elements, except for a small terminal region where each of the longer lateral elements pair homologously (Fig. 3a). Such an event could originate at early zygotene, when groups of telomeric ends frequently observed in close proximity could undergo mispairing (Fig. 3b). A further example of heterologous pairing always involves two SCs, which undergo an exchange of axial element (Fig. 3c) a short distance from one set of telomeres. That these are not translocation-type events is evident from their non-random subtelomeric localization. Such events appear to orginate at zygotene (Fig. 3d). Heterologous telomeres form a small pairing segment, while homologous pairing proceeds from a subtelomeric position. Finally, heterologous synapsis may involve the association of three lateral elements. While this might be expected in trisomic situations (Wallace and Hultén 1983; Speed 1984), it is perhaps surprising in karyotypically normal female foetuses. The three telomeres are most often in register (Fig. 3e); occasionally the associated non-homologue initiated heterologous synapsis away from the normally paired telomeres (Fig. 3f). Such triple synapsis appeared to originate at zygotene (Fig. 3g). The reasons why female germ cells should exhibit more synaptic errors than the male may relate to the greater length of the female SC. Bojko (1983) has estimated that the human female SC complement is about twice as long as those in spermatocytes contained within a nucleus of similar dimensions (Fig. 4a, b).

Intranuclear time schedules may mean that larger female SCs will find difficulty in reaching membrane synaptic initation points, and pair heterologously with the nearest available partner to satisfy synaptic drive.

3 Mutation and Meiosis

On exposure to either physical or chemical mutagens, an organism is likely to suffer mutations of varying types depending on the agent, the level of exposure, and the tissue involved, be it somatic or meiotic. Somatic effects are revealed as cytogenetic damage, with increasing levels of chromosome damage or sister chromatid exchange, and can usually be quantitatively related to dose. Damage to germ cells can induce gene mutations that may affect fertility or produce chromosome damage, such as translocations, which may be passed to future generations. Ionizing radiation has been the most frequently used experimental damaging agent. Chemical mutagens present a more complex situation with regard to the degree of penetration into the germ cell system, variable interaction with DNA, presence of de-activating enzymes and DNA repair systems etc. Meiotic mutation in mammals can be divided into several classes, each of which may affect the eventual reproductive potential of the individual – point mutations may influence genes controlling the normally functioning meiotic system in males and females.

Chromosomal mutation of a structural nature will include duplications and deficiencies, inversions and translocations. Cytological identification will depend on the size of segments involved. Genome mutations will involve changes in the number of chromosomes present, and can involve aneuploidy (hypoploidy, minus

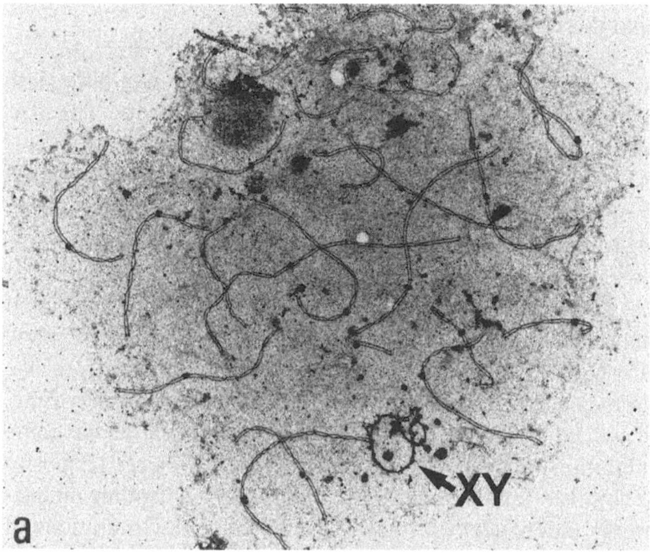

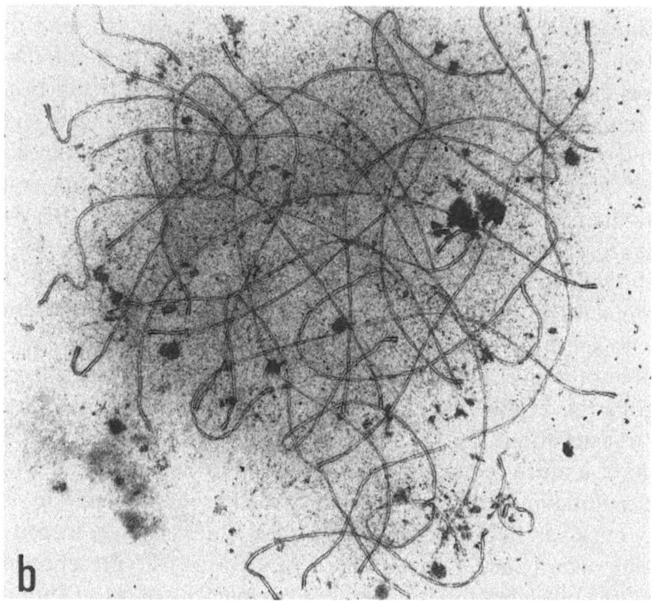

Fig. 4 a, b. Comparison of SC length in human spermatocytes and oocytes. **a** Pachytene spermatocyte with type II *XY* bivalent *(arrow)*. **b** Pachytene oocyte at same magnification, showing the total SC complement to be almost twice as long as in **a.** With permission of the Editor, Kew Chromosome Conference Proceedings, No III (Speed RM, Chandley AC p 313–321 1988)

one or more chromosome; hyperploidy, plus one or more chromosome) or polyploidy (an additional entire set, or sets, of chromosomes).

3.1 Gene Mutation and the Genetic Control of Meiosis

The normal progression of meiosis, as with all cellular processes, will ultimately be controlled by the genetic constitution of the organism (Rees 1961). The estimated number of approximately 50 000 gene loci present in the human genome is considerably in excess of the number of documented genes of known function either at the cellular or developmental level. Control of the highly specialized meiotic system leading to male and female germ cell production must account for at least some of these as yet undefined genes. The formation and control of specialized organelles, such as the synaptonemal complex, and the associated process of recombinational synapsis, will also involve the operation of many structural and regulatory genes specifically programmed to function at precise, but differing, times in males and females. Genes controlling the normally functioning meiotic system will be difficult to identify , and most information has come from the effects of mutation in species as diverse as *Saccharomyces* (Esposito 1984) and *Drosophila* (Lindsley and Sandler 1977). The genes controlling meiosis in *Drosophila* are largely independent in males and females, few mutations of such meiotic genes producing effects in both sexes. In addition, most data has come from the female as crossing-over is restricted to that sex. The mutant genes c3G and ord (Esposito 1984) function at the prepachytene stage and lead to abnormal SC formation, reduced recombination and increased levels of non-disjunction. In most cases the effects of such mutants are complex, and the function of the normal allele remains difficult to define at the level of gene products and function within the nucleus. Even in the mouse, the most widely studied experimental mammal, only about 25 genes are known to have a serious effect on fertility, which is less than half of those known to influence coat colour (Searle 1982). Such genes may affect early germ cell maturation, dominant spotting (W^v) causing a failure of the mitotic division in the primordial germ cells, and of their migration to the germinal ridge. The hop-sterile gene (hop) affects late maturation of the spermatozoa, no sperm tails being formed (Handel 1987). Genetic background can also be of importance; Forejt (1981) describes a locus male sterility-1 (Mst-1), which controls the extent of spermatogenic impairment in male mice carrying the T6 Ca translocation. On the inbred C57BL/6 background, T6 Ca heterozygote males have a much reduced testis weight and low sperm count. No positive examples of meiotic mutants have been described in the human, but as in other species, they must surely exist. Subfertility surveys occasionally identify individuals with pairing disturbance at metaphase I of meiosis. The presence of univalents, single chiasma bivalents and fragments in the majority of cells (Thomson et al. 1979) arise because of a reduction in chiasma frequency. Such males are oligo- or azoospermic and it has been proposed by Pearson et al. (1970) that an asynaptic or desynaptic mutation is the causal agent.

3.2 Meiotic Mutation Induced by External Agents

A large body of literature exists describing the effects of radiation and chemical mutagens on the somatic and germ cells of mammals and man. Historical reviews by Schleiermacher (1970) and more recently by Kimball (1987) cover the subject from Muller's (1928) first demonstration of X-rax induced mutation in *Drosophila*.

Many experimental approaches to dose response have involved exposure to mutagens with subsequent examination of levels of cell killing, or after a suitable delay (dependent on cell type), estimating levels of chromosomal damage in the form of translocations as seen at metaphase I or II of meiosis.

In the male mammal the late spermatogonia are the cells most sensitive to X-rays. Mouse B spermatogonia have an LD_{50} of only 21 cGy. Type A spermatogonia represent a mixed population, with those of low resistance only tolerating levels similar to type B, while the more resistant forms can survive 2–3 Gy (Oakberg 1957). Initial fertility following radiation is not affected at the above dose levels, as the later germ cell stages (spermatocytes, spermatids and spermatozoa) proceed with regular development. The most sensitive stage in the latter group is the spermatocyte with LD_{50}s ranging from 4–6 Gy (Oakberg and Di Minno 1960). Peak yields of radiation-induced translocations from spermatogonia occur at about 6 Gy, with a rapid return to fertility from repopulation by resistant gonia (Leenhouts and Chadwick 1981). Lower doses will produce the same levels of translocation in hamsters and rabbits (Lyon and Cox 1975), but return to fertility is slower. The response in various monkey species is variable (Van Buul 1984: Lyon et al. 1976; Matsuda et al. 1984) with a great delay in fertility recovery. Data for humans are limited. American male prisoners exposed to doses ranging from 7.5 cGy to 6 Gy in two separate studies were shown to recover fertility from all doses up to 4 Gy (Heller et al. 1975), while maximum delay for such recovery was 501 days (Thorslund and Paulsen 1972). Cawood and Breckon (1983) have recently extended the surface-spreading technique to male meiosis in Syrian hamsters to evaluate X-ray damage as measured at pachytene, in comparison with the traditional air-dried MI stage. They have shown the system to be a far more sensitive measure for structural rearrangements (Fig. 5a). Nearly twice the number of multivalents can be seen at the earlier stage of meiosis after an exposure of 2.6 Gy. Studies by Kaliakinskaya et al. (1986) have shown that in male mice exposed to 5 Gy ^{137}Cs rays, surface spreading in their F1 progeny permits identification of chromosome rearrangements not detactable at diakinesis. A further approach to the analysis of human spermatogonial damage by X-irradiation has entailed the use of the human sperm/hamster oocyte fertilization technique developed by Rudak et al. (1978). Baseline estimates of chromosomal abnormalities in normal males have been obtained by Martin et al. (1983). The same group (Martin et al. 1986) have recently shown that males receiving therapeutic testicular irradiation of up to 5 Gy had significantly increased levels of chromosomal rearrangements (21%) compared with controls (8–5%). Brewen et al. (1975) had already shown that translocation frequency was increased in human spermatogonia after X-ray exposure, this demonstrated that they were capable of transmission into the spermatozoa and that the results were significantly correlated with dose.

In the female meiotic system, germ cell sensitivity is again dependent on cell stage, dose received and species treated. As described earlier, the female mammal undergoes the prophase stage of meiosis before birth, and the adult has a fixed number of mature oocytes, with no capacity to replace any damaged by mutagens from a resistant stem cell pool. Effects of mutagens on female fertility may then be of greater consequence. Tateno and Mikamo (1984) have reported that murine foetal oocytes are radio-resistant from the prophase stages, leptotene to pachytene, 1 Gy causing no oocyte killing of the Chinese hamster. Progression to diplotene and the early dictyate stage at around the time of birth, however, brings a change to extreme radiosensitivity, 1 Gy killing the majority of oocytes within 48 h. Differences in chromosomal organization, with pachytene homologues being condensed and combined with proteins and RNA (in contrast to the more diffuse state seen at diplotene/dictyotene) are suggested as being responsible for the differences in X-ray sensitivity.

The adult mouse oocyte in the most immature follicle stage has an LD_{50} of only 8 cGy (Oakberg 1960), while the corresponding figure for rat and human are 1 and 20 Gy respectively. Oocytes in more mature follicles are considerably more resistant and LD_{100} values of 20, 40 and 50 Gy for mice, rats and humans have been reported by Gilliavod and Leonard (1973).

Chemicals ranging from anti-inflammatory drugs (Shobha Devi and Polasa 1987) to known mutagens (Röhrborn 1970), unlike X-rays, produce little germ cell damage in the form of chromosome structural rearrangements. It has been argued by Kimball (1987) that male cells with a long cell cycle, such as spermatogonia, will have the ability to repair DNA damage by agents such as MMS and EMS, before DNA replication takes place. However, in the adult female mouse, ewe and cow, follicular oocytes cultured in the presence of SO_2 and its metabolite (Na_2SO_3) show structural rearrangements at both metaphase I and II (Jagiello et al. 1975). Surface-spreading techniques may also be applicable in monitoring chemical damage at the oocyte pachytene stage. Allen et al. (1987) and Backer et al. (1988) have demonstrated considerable structural change in rodent SC morphology, following treatments with mitomycin C and cyclophosphamide. While such observations do not directly prove a relationship with subsequent chromosomal rearrangement seen at MI, it indicates that SC analysis is a sensitive measure of mutagen damage.

3.3 Meiosis and Chromosome Mutations Leading to Infertility

Chromosomal imbalance has been shown to have a strong correlation with impaired germ cell development in both mammals (De Boer 1986; Handel 1987) and man (Chandley 1979). In the former, experimental breeding protocols have allowed the development of a wide range of abnormalities that are open to experimental investigation. In man, the majority of data have come from male patients attending subfertility clinics. A figure of 5.3% for chromosomally abnormals, three times that of the newborn population, is documented from the four largest surveys undertaken (Kjessler 1974; Koulischer and Schoysman 1975;

Table 2. Chromosomally abnormal karyotypes found within the Edinburgh subfertility clinic

Karyotype	No.
Robertsonian translocation	4
Reciprocal translocation	10
47XXY	24
47XYY	5
46XY/47XYY	1
45X/48XYYY	1
45X/46r (Y)	1
46X inv(Y)(p11;q11)	1
47XY, mar+	4
Total	51

Chandley 1979; Tiepolo et al. 1981). It also became evident that as the sperm count was lowered, the number of chromosomal abnormalities substantially increased. From the Edinburgh subfertility survey of 2375 males, for those with a sperm count in the range 21–60 $\times$ 10^6/ml, only 0,94% abnormalities were recorded, but with a sperm count reduced to 1 $\times$ 10^6/ml or less, abnormalities rose to a high of 15.38% (Chandley 1979). The composition of chromosomal abnormalities found at Edinburgh are given in Table 2. It is clear that sex chromosome abnormalities are strongly represented, particularly the 47XXY class, these falling within the severe oligospermic to azoospermic range. Translocations of both the Robertsonian and reciprocal type were recorded as they have been consistently so in other surveys.

Investigation of subfertile adult human females to determine the influence of chromosome abnormalities, already shown to be of importance in male subfertility, have been limited. Of some 850 women at an Edinburgh subfertility clinic, only five were found to be karyotypically abnormal (Jacobs et al., unpublished data, 1972). Three reciprocal and one Robertsonian translocation, plus one extra marker chromosome were identified. The resulting frequency of 0.59% abnormal was not significantly different from the 0.38% obtained amongst normal female controls. The nature and effects of these chromosomal abnormalities on subfertility as seen at the light microscope level have been extremely well reviewed elsewhere (Chandley 1984) and only more recent evidence from electron microscope studies will subsequently be dealt with.

3.3.1 Translocations

Experimental work, particularly with mouse translocations (reviewed by Gropp et al. 1982; Searle 1982), and studies of spermatogenic breakdown in human male translocation heterozygotes (reviewed by Chandley 1984) have greatly contributed to our understanding of mammalian germ cell death. Initially, the interactions of the translocations themselves within the nuclei, for example with the sex bivalent, may produce disturbance and cell death. Secondly, irregular segregation from translocations will create variable numbers of unbalanced gametes, contributing to foetal wastage, and an apparent reduced fertility. Problems of synapsis fall within the first category, and several hypotheses to explain the relationship of fertility and irregularities of pairing in both human balanced reciprocal translocations (involving combinations of autosomal acrocentrics, non-acrocentrics and sex chromosomes), and Robertsonian translocations (centric fusion between acrocentric autosomes) will be examined.

3.3.1.1 Autosome-Sex Chromosome Translocations

Both X and Y chromosomes may be involved in mammalian translocations, but the condition is rare. For the mouse this form of exchange (Fig. 5b) inevitably leads to sterility, with breakdown at pachytene frequently occurring (reviewed by Handel 1987). Madan et al. (1981) reviewed 14 human male carriers of differing X-autosome translocations, showing that all the adult carriers were oligo- or azoospermic, germ cell arrest at the primary spermatocyte level being the most common defect. A case involving an X-2 human translocation with associated azoospermia was recently examined in our laboratory using surface spreading (Quack et al. 1988). The translocation chromosomes at pachytene formed a chain quadrivalent in 86% of cells. Even though the X and Y chromosomes were involved in the quadrivalent, the sex bivalent type could still be assessed (Fig. 5c), only types I and II being observed; this indicated an early arrest. While asynapsis was common in the vicinity of the breakpoints, approximately 18% of the quadrivalents showed heterologous pairing in the central region. Ashley et al. (1983) reported similar findings in two mouse X-7 translocations. It appears that sections of the X chromosome, lying outside the normal region that synapses with the Y chromosome, can heterologously pair with lateral elements of an autosomal origin. The fertility of human female carriers will depend on where the breakpoints occur. Those located in the "X-critical" region (Xq13-Xq26) will lead to primary and secondary amenorrhoea (Sarto et al. 1973). Human Y-autosome transloca-

Fig. 5 a–g. Mammalian reciprocal translocations at the electron microscope level. **a** X-ray induced translocation, t(11L;20L) in a male Syrian hamster *(Mesocricetus auratus)*. With permission of A.H. Cawood and G. Breckon (unpublished data). **b** Mouse, T38H, X-autosome translocation, t(X;11). **c** Human, X-autosome translocation, t(X;2), Y and portion of X at pairing stage II *(arrowhead)*. **d–h** Mouse T(14;15) 6Ca translocation. **d** Fully paired quadrivalent showing heterologous synapsis at breakpoints *(arrow)*. **e** Sections of chromosomes 15 and 15^t showing a thickened morphology, subsequent to pairing failure in a short arm. **f** X chromosome showing contact with fully paired quadrivalent *(arrow)*. **g** Trivalent (chromosomes 14, 14^t and 15) showing heterologous short arm synapsis *(arrow)*. **h** Trivalent in contact with X chromosome *(arrow)*. Univalent chromosome 15^t, showing thickened morphology and telomere association *(arrow)*. **d–h** Reprinted with permission from Fertility and Chromosome Pairing: Recent Studies in Plants and Animals. Copyright CRC Press Inc., Boca Raton, FL (Speed RM in press)

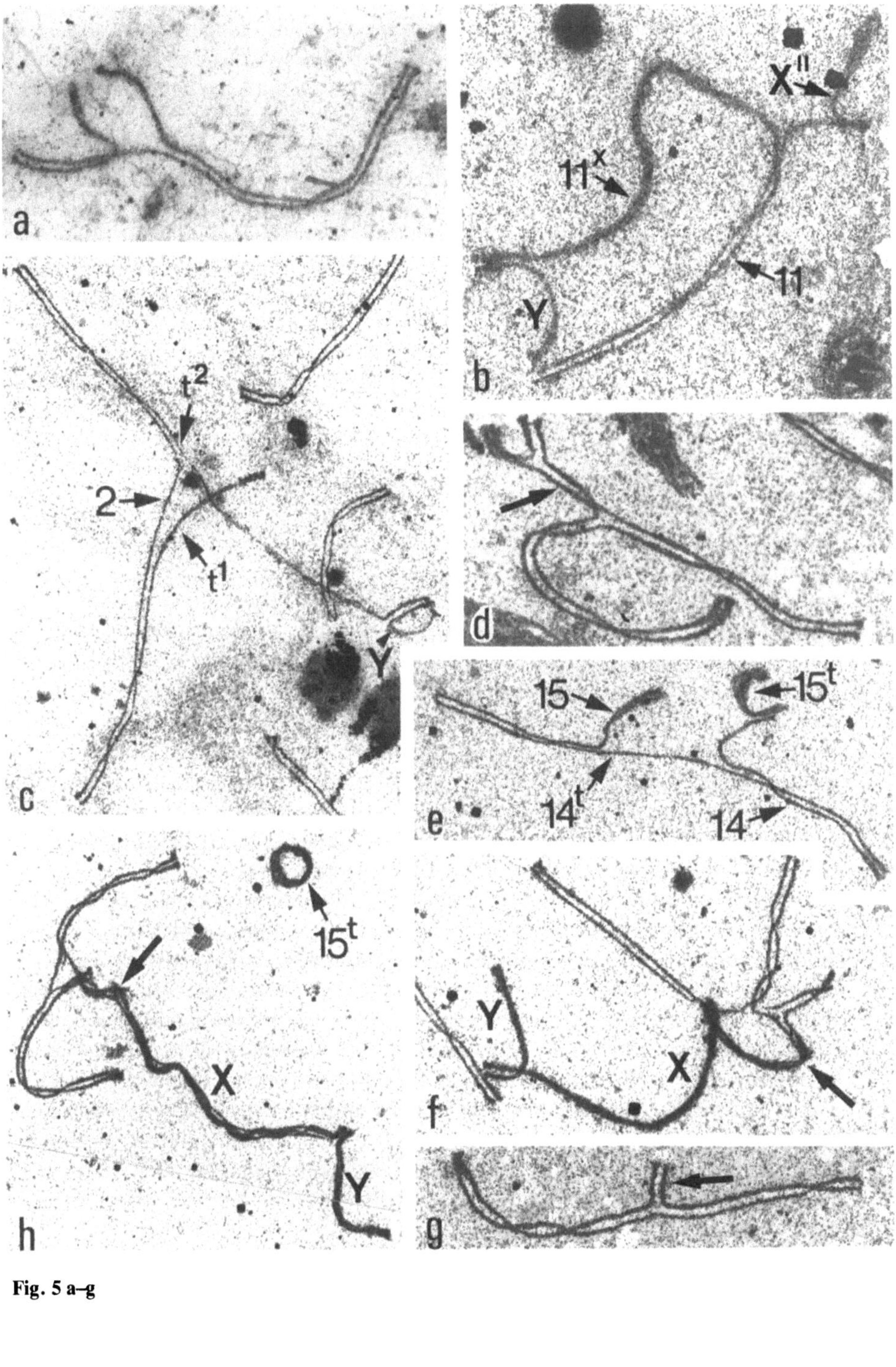

Fig. 5 a–g

tions, not involving transference of heterochromatic portions of the Y long arm to D and G group acrocentrics, lead to reduced fertility ranging from oligo- to azoospermia (Smith et al. 1979). Pairing disturbances in the form of univalents and unequal bivalents have been seen at metaphase I with the light microscope, but little information from surface spreading is as yet available.

A mechanism to explain the reduced fertility in the X/autosome situation has been proposed by Lifschytz and Lindsley (1972), who found that 80% of such translocation carriers in *Drosophila* were sterile. For normal spermatogenesis, it was deemed essential that pairing and transcriptional activity in the XY bivalent progressed asynchronously in relation to the autosome throughout prophase. Any disturbance in this relationship, such as the contact of the X with an autosome via a translocation, would lead to maturation arrest. Extended to mouse and man, if the normally inactive X were to become transcriptionally active at an inappropriate time, disturbances in germ cell maturation would lead to eventual infertility.

3.3.1.2 Autosome-Autosome Translocations Reciprocal

Translocations involving only autosomes also cause sterility in mammals. In the mouse, Lyon and Meredith (1966) observed that excess numbers of translocations with chain configurations at metaphase I of meiosis were associated with infertility. Further analysis has shown that the location of one breakpoint near a telomere was the causal factor: this produces an asymmetrical quadrivalent, with at least one short arm. Failure of synapsis within this short arm at pachytene will then lead to a chain configuration at MI. From 11 human reciprocal translocations reviewed by Chandley (1984), one azoospermic male had 100% of chains, while three males with sperm counts of 20×10^6/ml or more had no chains, and 100% ring configurations. Forejt (1982) has proposed, as an extension to the Lifschytz and Lindsley (1972) model, that quadrivalent contact with the X chromosome, causing its reactivation, might be implicated in germ cell breakdown. The initial observation by Forejt (1974) in the T(14;15)6 Ca mouse translocation at MI, however, indicated that it was the autosomal region in contact with the X chromosome that became heteropycnotic (transcriptionally inactive) and dark staining, resembling the X itself. Only in sterile male mice with double Robertsonian translocations did autosome/sex bivalent contact result in decondensation of the X and Y chromosomes (Forejt 1979). A preliminary investigation of T(6) Ca male mice using surface spreading at prophase has shown a more complex pattern of events occurring, several of which may lead to cell death (Speed RM, unpublished observations, 1988). The quadrivalent is only occasionally fully paired, and often shows heterologous synapsis in the region of the breakpoints (Fig. 5d). Failure of synapsis and thickened morphology in the arm concerned occurs prior to association with the XY bivalent (Fig. 5e). The quadrivalent may also associate with the XY bivalent with no apparent asynapsis or thickening of autosomal material (Fig. 5f). The small marker 15^t chromosome may also fail to pair, remaining as an univalent. The remaining three translocation chromosomes then pair as a trivalent-like structure, with heterologous synapsis occurring between chromosomes 14 and

Table 3. Human reciprocal translocations, their possible association with the XY bivalent and its subsequent relationship to fertility

Karyotype	Association of quadrivalent with XY (%)	Sperm count or testicular histology	Reference
Not involving acrocentrics			
46,XY,t(4;17)	0.00	45 × 10^6/ml	Speed unpublished data (1988)
46,XY,t(2;4;9)	1.20	122 × 10^6/ml	Saadallah and Hultén (1985)
46,XY,t(4;16)	1.60	Normospermia	Batanian and Hultén pers comm (1987)
46,XY,t(7;20)	5.20	Reduced spermatogenesis	Batanian and Hultén pers comm (1987)
46,XY,t(9;20)	20.00	9.2 × 10^6/ml	Chandley et al. (1986)
46,XY,t(10;17)	31.00	Maturation arrest	Hultén et al. (1987)
Involving acrocentrics			
46,XY,t(9;12,13)	0.00	Normospermia	Johannisson et al. (1988)
46,XY,t(9;21)	11.00	27.9 × 10^6/ml	Batanian and Hultén pers comm (1987)
46,XY,t(9;15)	42.50	32.6 × 10^6/ml	Johannisson et al. (1987)
46,XY,t(14;21)	65.50	Azoospermia	Johannisson et al. (1987)
46,XY,t(17;21) 46,XY,t(19;22)	70.00	Azoospermia	Gabriel-Robez et al. (1986a)
46,XY,t(17;21)	77.70	3.9 × 10^6/ml	Luciani et al. (1987)

15 (Fig. 5g). Such trivalent-like structures may associate with the XY bivalent (Fig. 5h) as do the partially asynaptic quadrivalents. Obviously a more complex synaptic pattern of autosome/XY contact can exist at prophase compared with that seen at metaphase I.

The hypothesis of Miklos concerning unsaturated pairing sites could then also be extended to translocations. Asynaptic regions are common in unpaired side arms and around the breakpoints involved, as above in T(6) Ca, and in human translocations. Further, the smallest chromosome involved in the T(6) Ca translocation may act as an univalent, being included in the sex vesicle as in tertiary trisomic mice (de Boer and Brange 1979) where subfertility is usual.

Human reciprocal translocations can be divided into two groups. The first does not involve the acrocentric D and G group chromosomes, while the second does and exhibits increased germ cell death. The former group in general shows lower numbers of quadrivalent/XY association and higher sperm counts (Table 3). General pairing disturbances are seen, as in the 46,XY,t(9;20) studied at Edinburgh (Chandley et al. 1986). Asynapsis around the breakpoints (Fig. 6a) is a common feature, and in the few quadrivalents that do fully pair, heterologous synapsis occurs in the central region comparable with that seen in the mouse T(6) Ca translocation.

Again, the main causal agent promoting contact between the quadrivalent and XY bivalent is failure of synapsis in one arm of the former, this occurring in about 20% of the cells. The morphology of the unpaired autosomal-lateral elements is again similar to the X and Y that they contact (Fig. 6b). There also appeared to be large numbers of germ cells degenerating before they entered the

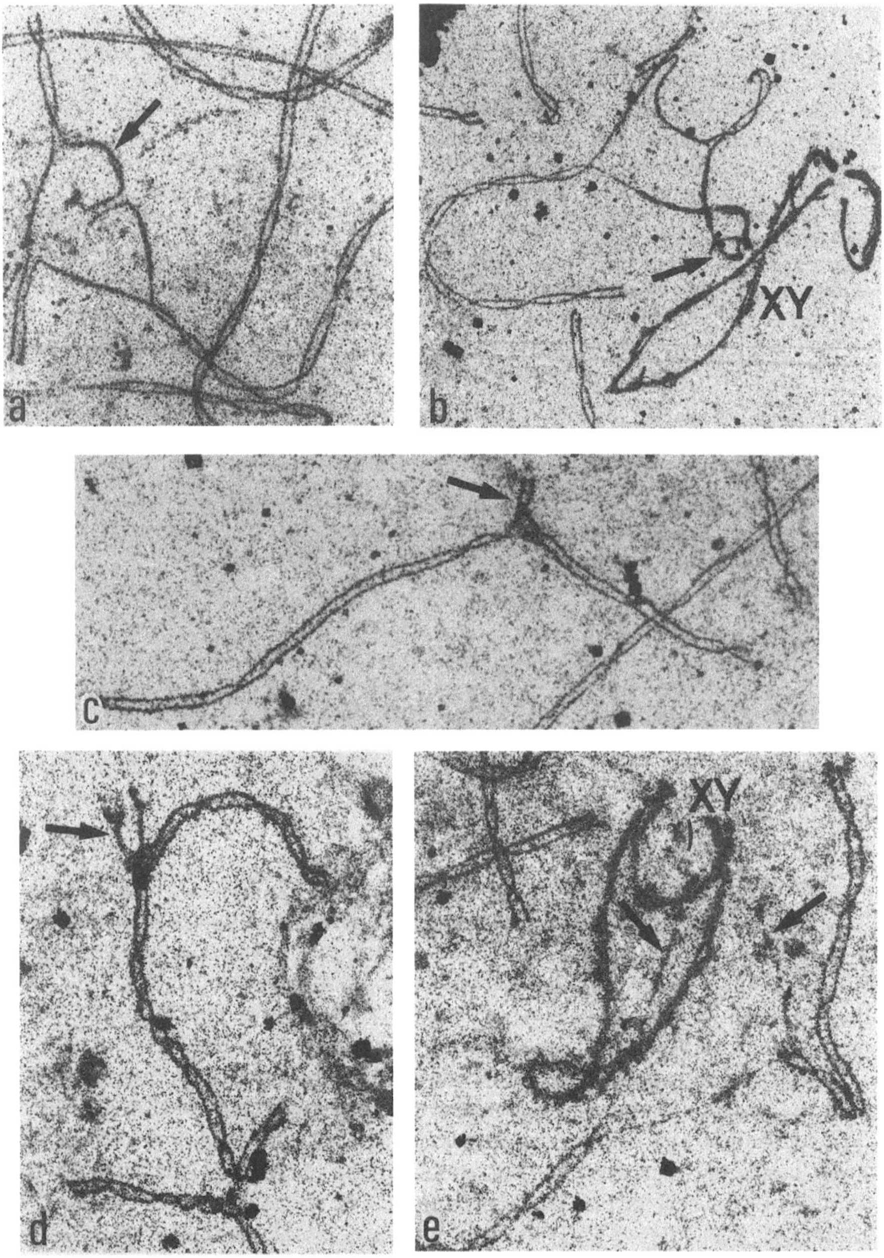

Fig. 6 a–e. Human reciprocal translocations. **a** A t(9;20) quadrivalent showing pairing failure in the region of the breakpoints *(arrow)*. **b** A t(9;20) quadrivalent showing association between a thickened, unpaired, short arm region and the XY bivalent *(arrow)*. **c** A Robertsonian t(14;22) trivalent showing heterologous, short arm synapsis *(arrow)*. **d** Robertsonian trivalent with the short arms of chromosome 14 and 22 unpaired *(arrow)*. **e** Unpaired short arms *(arrow)* of t(14;22) trivalent in association with the XY bivalent. **a, b** With permission of Cytogenetics Cell Genet (Chandley et al. 1986). **c–e** Reprinted with permission from Fertility and Chromosome Pairing: Recent Studies in Plants and Animals. Copyright CRC Press Inc, Boca Raton, FL (Speed RM in press)

prophase of meiosis, suggesting that translocation-associated spermatogenic breakdown may even be initiated premeiotically. Gabriel-Robez et al. (1986a) have shown that when acrocentric chromosomes become involved in human translocations, as compared to translocations with no acrocentric involvement, that a ratio of 1:4 of such rearrangements in fertile males changes to1:1 in males attending subfertility clinics. The heterochromatic, NOR(nucleolus organizing region)-bearing short arm regions of acrocentric chromosomes in normal males have been shown to associate with the sex vesicle (Stahl et al. 1984) and this association can also be seen in spread preparations (Fig. 2d). Does the presence of acrocentric regions then add to the affinity of asynaptic regions in translocations enhancing cell disruption? The available human data shown in Table 3 is limited, but is suggestive of increased levels of fertility disturbance, reflected in lowered sperm counts.

Female carriers of reciprocal translocations have previously been thought of as fully fertile. However, the female can be effectively fertile, with far fewer germ cells than is the case in the male. Mittwoch et al. (1981) have shown that the mouse T42H translocation, causing male sterility, also causes impairment of female germ cell development, the numbers of oocytes and ovary size 1 week after birth being much reduced. Little evidence exists for humans, but in the aneuploid XO human foetal ovaries, a premeiotic block exists, with few oocytes progressing to maturity.

3.3.1.3 Robertsonian Translocations

Robertsonian translocations for all the mouse chromosomes have been isolated, occasionally from laboratory stocks, but in the main from feral groups in Central Europe. Their use in non-disjunctional and fertility studies has been extensively reviewed by Gropp et al. (1982). Human Robertsonian translocations comprise the largest group of translocation heterozygotes detected at infertility clinics (Tiepolo et al. 1981) and usually involve chromosomes 13 and 14. The effects

Table 4. Human Robertsonian translocations: their possible associations with the XY bivalent and subsequent Relationship to fertility

Karyotype	Association of quadrivalent with XY (%)	Sperm count or testicular histology	Reference
45,XY,t(13;14)	0.0	27.6×10^6/ml	Templado et al. (1984)
45,XY,t(13;14)	0.0	22.4×10^6/ml	Templado et al. (1984)
45,XY,t(13;14)	5.3	14.0×10^6/ml	Batanian and Hultén pers. comm. (1987)
45,XY,t(13;14)	8.8	Azoospermia	Batanian and Hultén pers. comm. (1987)
45,XY,t(14;22)	12.0	15.0×10^6/ml	Speed unpublished data (1988)
45,XY,t(13;14)	61.0	1.6×10^6/ml	Luciani et al. (1984)
45,XY,t(14;21) 45,XY,t(14;21)	75.0	Severe oligospermia	Rosenmann et al. (1985)

of such centric fusions on human fertility vary from azoospermia to normal (Table 4). The fertility of individuals within families carrying the same translocated chromosomes may be variable and, as Rosenmann et al. (1985) have pointed out, such carriers will inherit different combinations of the normal chromosomes 13 and 14, which combine with the centric fusion chromosomes to form the trivalent at meiosis. Genetic background may then also be of importance in such rearrangements. Variable contact with the XY bivalent ranging from 0% (Templado et al. 1984) to 75% (Rosenmann et al. 1985) is observed at pachytene, reduced fertility occurring at the upper limits (Table 4). Trivalents can be fully paired (Fig. 6 c) or as in a 45,XY,t(14;22) patient studied in Edinburgh, 57.6% of trivalents showed the acrocentric short arms asynapsed (Fig. 6d). This category may (21.1%) or may not associate with the XY bivalent (Fig. 6e).

Again in Robertsonian translocations, as with the reciprocal variety, do asynaptic telomeric regions themselves, or their possible subsequent contact with the XY bivalent, constitute the major defect leading to cell death and reduced fertility? The situation has recently been reviewed by Handel (1987). While the Miklos hypothesis of unsaturated pairing sites in univalents or asynaptic regions can satisfactorily explain certain situations, nothing is kown as to the nature of such sites in causing germ cell degeneration. The problem of quadrivalent or trivalent contact with the XY causing a non-permissible reactivation of the X and infertility as proposed by Forejt (1982) must also be reconsidered. This mechanism of sterility originating from the observation of *Drosophila* X/autosome translocations, by Lifschytz and Lindsley (1972), was thought to be due to the autosome activity transmitting past the breakpoints and into the attached part of the X chromosome. Hackstein (1987), however, has recently stated that the cytological evidence for precocious X inactivation in primary spermatocytes of *D. melanogaster* is poor, and that Kremer et al. (1986) using the DNA dye DAP1 found no evidence of X inactivation during premeiotic development of *D. hydei*. Further, of 600 X-chromosome *Drosophila* male steriles, some mutations caused abnormal spermatocyte development, suggesting their activity rather than inactivity at this stage (Lifschytz and Hareven 1977). Translocations also appear to induce premeiotic breakdown, and Hackstein (1987) argues that the possibility of interference with dosage compensation, caused by overactivity of autosomal genes near to the breakpoints, might disturb germ cell development. Obviously, at the pachytene level further advances in the biochemical nature of events are required for a clearer understanding of the underlying problems.

3.3.2 Inversions

Pericentric inversions in mammals would be expected to reduce fertility by the production of unbalanced gametes (duplication deficiency) due to crossing-over within the inverted segment. Ascertainment in human has been through the identification of children born with congenital malformations. However, studies of synaptic behaviour at prophase and metaphase I of meiosis have been possible in the male because of material obtained from subfertility clinics. Human chro-

Table 5. Human inversions, heterologous pairing and fertility

Karyotype	Cells showing heterologous pairing within inversion region (%)	INV-XY association (%)	Sperm count or testicular histology	Reference
45,XY,Inv(13)(p12q14)	74.2	11.0	Normal	Saadallah and Hultén (1986)
[a]46,XY,Inv(1)(p32q21)	22.6	0.0	6.14×10^6/ml	Guichaoua et al. (1985)
[a]46,XY,Inv(9)(p11q12)	27.4	0.0	6.14×10^6/ml	Guichaoua et al. (1985)
46,XY,Inv(1)(p32q12)	16.7	0.0	Oligospermia 1.2×10^6/ml	Gabriel-Robez et al. (1986b)
46,XY,Inv(1)(p31q43)	16.4	0.0	Oligospermia 1.0×10^6/ml	Chandley et al. (1987)
46,XY,Inv(1)(p32q42)	0.0	0.0	Azoospermia	Batanian and Hultén (1987)

[a]A single male heterozygous for two pericentric inversions.

mosome number one appears to be most frequently associated in the reduced germ cell development (Table 5). Surface-spread spermatocytes have demonstrated an apparent rarity of classical fully synapsed loop formation, in both human and animal species (Chandley et al. 1987). Instead, the inverted region may show partial loop formation, or extensive asynapsis, with the typical thickened and dark staining appearance of pairing failure (Fig. 7a and b). Finally, the inverted region may synapse heterologously through its length, forming an SC indistinguishable from normal.

In mice heterozygous for tandem duplications, or paracentric inversions, a phenomenon of synaptic adjustment has been described (Moses and Poorman 1981; Moses et al. 1982). Fully synapsed inversion loops decrease in size from early pachytene, until by mild/late pachytene the loops have resolved themselves by secondary heterologous pairing and resemble normal SCs. Heterologous pairing in some human inversion cases has been reported along the inversion at the late prophase stage when synaptic adjustment normally occurs (Guichaoua et al. 1985; Gabriel-Robez et al. 1986b). However, in the Inv(1) studied by Chandley et al. (1987), fully heterosynapsed inversion bivalents were seen in the earliest type 0 spermatocytes. This was also the case in a patient with an Inv(13) reported by Saadallah and Hultén (1986). Such an early zygotene initiation of heterologous synapsis has been observed by Hale (1986) in *Peromyscus sitkensis* (deer mouse), which may also be a general characteristic of this species. Little evidence of contact between the inversion asynaptic region and the XY bivalent has been observed in man (Table 5) apart from the Inv(13) case (Saadallah and Hultén 1986). Contact with the sex bivalent, as previously discussed, usually involves unpaired telomeric regions of chromosomes, as present in univalents or asynaptic arms of translocation quadrivalents; infertility as proposed by Forejt is unlikely in patients carrying

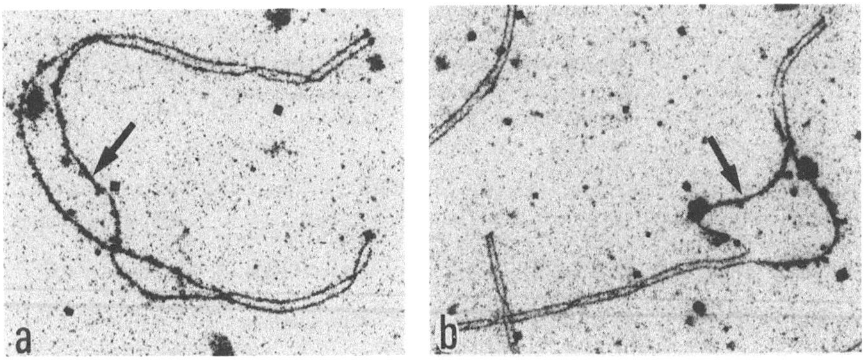

Fig. 7 a, b. Human inversion of chromosome No. 1. **a** Extensive interstitial asynapsis within the inverted region *(arrow)*. **b** Progressive heterologous synapsis, has reduced the extent of a synapsis in the inversion region. With permission of J Med Genet (Chandley et al. 1987)

the inversion. A further general asynapsis of both autosomes and sex chromosomes was seen (Batanian and Hultén 1987; Chandley et al. 1987) in two inversion carriers. Such an interchromosomal effect might lead to cell death according to the theory of Miklos. If heterologous pairing is frequently seen in inversion carriers, classical crossing-over within an inversion loop may be less frequent than previously thought. In the case reported by Chandley et al. (1987), reduced chiasma frequency both within the inversion bivalent and specifically within the inversion segment was evident. It may be that fewer unbalanced recombinant segregants than expected will be seen amongst the offspring of inversion carriers.

3.4 Meiotic Genome Mutation and Infertility

3.4.1 Autosomal Trisomy

Mammalian trisomies usually exhibit considerable cellular disturbance such that few survive to birth. All the mouse trisomies have been examined by manipulation of the Robertsonian translocations found in naturally occurring populations of wild mice (Gropp and Winking 1981; Epstein 1985). Only trisomy 19 survives for a short period postnatally. Examples in other mammals are occasionally recovered, ranging from the horse (Power 1987) to the orangutan (Andrle et al. 1979), but since few survive to maturity the effects on fertility are undocumented, except in a trisomy 22 cow (Mayer et al. 1987) which gave birth to both chromosomally normal and abnormal calves. Only the human trisomy 21 survives into adulthood. Sperm counts in nine such cases showed four to be azoospermic, the remainder having low counts in the subfertile range (Stearn et al. 1960). Johannisson et al. (1983) showed that the extra chromosome 21 at metaphase I can remain as an univalent (88.5% of nuclei) or form a trivalent (4.8% of nuclei): the remaining cells appeared to have lost the extra chromosome. Differing ratios seen at pachytene (8.5%, univalent; 2.4%, trivalent) can be explained if the extra chromosome 21 is included in the sex vesicle with the X and Y chromosomes. Their tangled morphology at the late stages of pachytene make the inclusion of an extra element difficult to visualize. In a similar situation analysis of tertiary trisomic male mice (derived from the T70H translocation) by De Boer and Branje (1979) shows that the extra chromosome is almost exclusively contained in the sex vesicle. In both mouse and man, the Lifschytz and Lindsley model has been used to argue that the presence of the autosomal material in the sex vesicle disrupts normal spermatogenesis.

Fertility is possible in Down's females, there being 27 recorded cases of live births. The expected 1:1 segregational pattern is closely approximated by the observed ratio of normal to trisomy 21 offspring (Jagiello 1978). A study of trisomy 21 foetal oocytes at the EM level (Speed 1984) has shown that as in the male, the three chromosomes 21 can synapse in a triple association (Fig. 8a). Compared with the male, the trivalent configuration is more frequently observed (33.8% cf 2.4%) than the free univalent (60.0% cf 8.5%). The apparent synaptic success of the third chromosome 21 in the female merely reflects the absence of an isolated partially asynaptic female sex bivalent. The extra chromosome 21 in the female only has the option to triple pair, or remain as an identifiable univalent.

Trisomy can also be caused by the presence of translocations or isochromosomes. Foetal ovaries from both a 46,XX,G-,t(21;21) and a mosaic 46,XX 19p-; iso 18q foetus studied at Edinburgh have shown similar synaptic behaviour. In the former triple pairing between chromosomes 21 involved in the Robertsonian translocation, and the normal 21 (Fig. 8b) occurred in only 19.2% of the nuclei. In the foetus with the iso 18 cell line, triple pairing (Fig. 8c) occurred with a much higher frequency in 68% of the nuclei.

A feature of both these foetuses was an apparent meiotic delay in oocytes reaching the pachytene stage, as compared with chromosomally normal foetuses of the same gestational age. Germ cells appeared to block at a preleptotene stage. Such an effect may account for the severe reduction in germ cell numbers seen in the ovaries of surviving trisomy 18 babies (Russell and Altschuler 1975) and trisomy 21 girls (Peters et al. 1981). In the foetal testes of the trisomy 13, 18 and 21 condition, severe reductions of germ cell numbers also occur (Coerdt et al. 1985). Delay and degeneration thus appear to be characteristic of human aneuploid germ cell development in both males and females, the earliest onset being seen in the foetal gonads.

3.4.2 Sex Chromosome Aneuploidy

The effects of sex chromosomal changes in mammals have been reviewed in general by Russell (1976), for mice by Searle (1982) and domestic farm animals by Blazak (1987). The most common human sex chromosome aneuploidy is the 47,XXY Klinefelter's syndrome. Such males are generally azoospermic with small testes, lacking any germ cell development, which is the general condition seen in

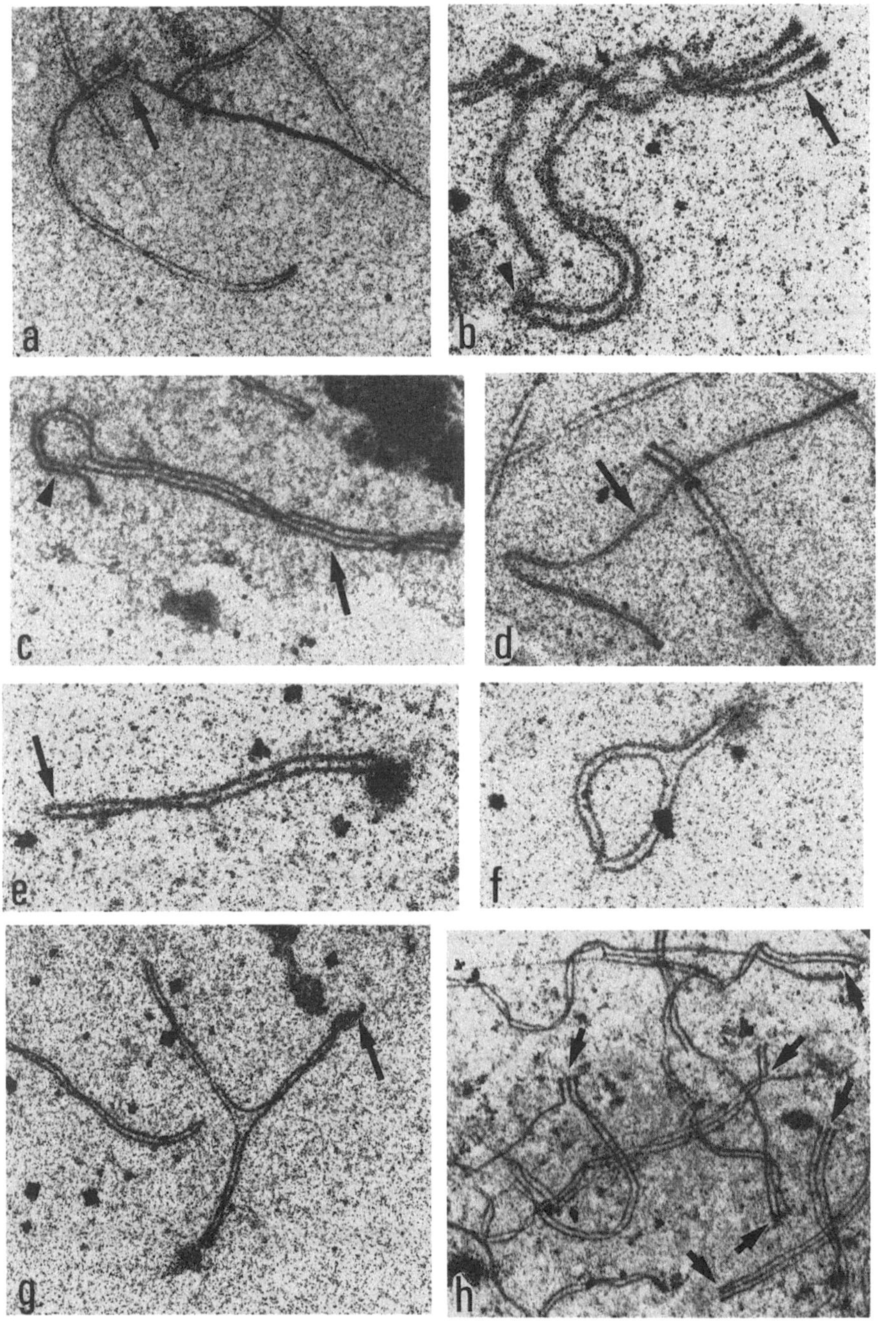

Fig. 8 a–h

most mammals. This situation is thought to arise because of the failure of germ cells with two X chromosomes to survive in a testicular environment.

In 47,XYY males fertility is extremely varied, ranging from virtual normality to almost azoospermia (Skakkebaek et al. 1973). The majority of meiotic data for XYY males suggest that only germ cells of an XY consitution are seen at metaphase I (Evans et al. 1970; Chandley et al. 1976). Other studies of sectioned human spermatocytes examined at the EM level have argued for the retention of the second Y until the pachytene stage (Berthelsen et al. 1981) and at metaphase I (Hultén and Pearson 1971), as in the case of the mouse. The XO chromosomal constitution is the most frequently occurring human female form of sex chromosome aneuploidy, with levels of 20 % seen in spontaneous abortion surveys (Chandley 1984). Those that survive to term and adulthood represent only 5 % of the numbers of XO conceptions. The condition is not so lethal in the mouse, where two-thirds survive to term (Brook 1983).

Most adult human XO females show a failure of ovarian development, even though foetal ovaries have been shown to contain germ cells (Carr et al. 1968). XO mice were initially thought to have normal fertility (Cattanach 1962), but subsequent studies by Lyon and Hawker (1973) have shown that they too have fewer oocytes than normal and a shortened reproductive lifespan. Burgoyne and Baker (1985) suggested that the presence of the univalent X chromosome might be the causal agent of such excess atresia in pachytene oocytes. Surface-spreading studies (Speed 1986b) investigating the fertility difference have shown that in human XO foetal ovaries, a block in progression occurs similar to that seen in autosomal trisomies. The majority of oocytes remain in a preleptotene stage. Those few that do progress to pachytene contain a single thickened axial element, interpreted as an X univalent (Fig. 8d). In the mouse, a minor delay in progression occurs, but the great majority of oocytes pass through pachytene and reach the dictyate stage by birth. The mouse X chromosome may also in a proportion of oocytes remain single and thickened, as in the human. The X chromosome in most oocytes, however, undergoes heterologous pairing. This may take the form of self-synapsis, forming hairpins or loops (Fig. 8e and f); alternatively, each telomere of the X can initiate heterologous synapsis with two homologous, but as yet unpaired, autosomal telomeres and synapse to form a triradial structure

Fig. 8 a–h. Synaptic events at the EM level in mammalian aneuploid oocytes. **a** Small region of terminal triple pairing *(arrow)* between the three chromosomes No 21 of a standard Down's syndrome foetus. **b** Triple pairing *(arrow)* in a 46,XX,t(21;21) Down's foetus. The Robertsonian translocation *(arrowhead)* has paired both with itself and the normal chromosome No 21. **c** Virtually complete triple pairing of a normal human chromosome No. 18 *(arrow)*, with an iso-18q. Homologous synapsis occurs almost to the centromeric region, with subsequent heterologous pairing between the short arm of the normal chromosome No 18 and the iso-18q *(arrowhead)*. **d** Single thickened human X axial element from an XO foetus. **e–g** Single X axial element from mouse XO foetal oocytes paired heterologously as **e** a hairpin *(arrow)*; **f** a ring; **g** with both axial elements of an autosome, forming a triradial structure. **h** Six regions of triple pairing *(arrows)* in a human triploid foetal oocyte. **a, c** With permission of Hum Genet (Speed 1984, 1986a); **e–g** With permission of Chromosoma (Speed 1986b)

(Fig. 8g). On day 19 of gestation, just before birth, approximately two-thirds of oocytes contained an X chromosome which, although lacking a homologous pairing partner, was able to satisfy its pairing instincts by heterologous synapsis. Survival of oocytes in the XO mouse ovary might therefore depend on such pairing behaviour, and would be in accord with the hypothesis of Miklos where "pairing sites" must be saturated to permit a regular maturation of germ cells. In the few cases of apparently fertile human XO women, it may be that the single X chromosome can also occasionally undergo heterologous synapsis, allowing oocyte survival into the adult ovary.

3.4.3 Triploids

Human triploidy occurs with a frequency of about 1 % in all clinically recognizable pregnancies. However, the majority of such conceptuses spontaneously abort; Jacobs et al. (1982) state that only about 1 in 10000 survive to term as abnormal infants, who normally die in the postnatal period. The question of fertility is therefore of little consequence, as would be the study of meiosis in the male, which would not commence until early adulthood. However, foetal ovaries obtained from several triploid foetuses have shown some meiotic development. Air-dried preparations (Gosden et al. 1976) showed that all prophase stages up to diplotene were present. More recently surface-spreading (Speed RM, unpublished observations 1988) has shown that multiple regions of triple pairing can be observed at pachytene (Fig. 8h), and that a delay in germ cell numbers reaching pachytene occurs when compared with chromosomally normal foetuses of the same gestational age.

4 Non-Disjunction and Aneuploidy

A wide literature on the origins and mechanisms of meiotic non-disjunction, leading to aneuploidy (spontaneous and induced) has accumulated over the last 2 decades. It is not the intention to deal in great depth with the subject, as its effects on mammalian fertility have already been examined. Several comprehensive reviews exist, including those by Bond and Chandley (1983), Dellarco et al. (1985) and a compilation by the leading workers in the field, edited by Vig and Sandberg (1987). The most common factor influencing levels of non-disjunction at MI of meiosis is increased maternal age, as first described by Penrose (1933). Mechanisms to explain the increased aneuploidy in older females have varied from nucleolar persistence (Evans 1967), delayed fertilization (German 1968), declining chiasma frequency (Henderson and Edwards 1968) and hormonal imbalance (Lyon and Hawker 1973).

Of the several approaches to aneuploidy studied at our laboratory, one has involved the investigation of the Henderson and Edwards (1968) "Production Line" theory. Their observations of reduced chiasma frequency and increased

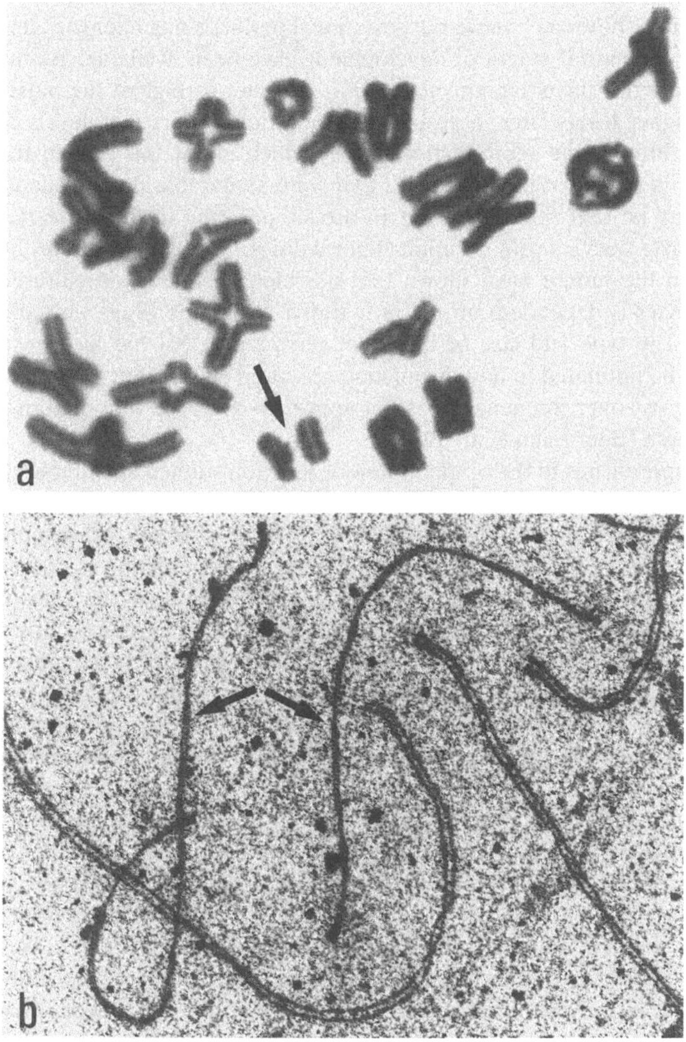

Fig. 9 a, b. Univalents in mammalian female germ cells. **a** An adult male mouse oocyte cultured for 48 h showing two univalents *(arrow)* at MI of meiosis (light microscope). **b** Two thickened axial elements (univalents) from a human foetal oocyte (electron microscope level)

univalent numbers (Fig. 9a), in the oocytes of aged female mice compared with those of young females, led to the suggestion that gradients of oocytes developed in the foetal ovary. Oocytes formed early in gestation with higher chiasma frequency, and few, if any, univalents would be ovulated sequentially early in the adult reproductive phase. Those with synaptic errors, formed last, would be ovulated late in life, contributing to the high levels of aneuploidy seen in older mothers. In the mouse and the human, oocytes of the foetal ovary are open to investigation by surface-spreading, and in the mouse, no evidence for increased univalent formation during pachytene has been observed (Speed and Chandley 1983). In the human, however, where the gestational period is much longer, only foetal ovaries of the earlier stages of development have been available. As the levels of synaptic error, including univalents (Fig. 9b), are so high at the outset (Speed 1988), higher levels later in gestation should only reflect high levels of non-disjunction through the reproductive phase, which is not the case in the human. Further, in the few diplotene stages that were analyzable, no univalents were seen; it may be that the high levels of atresia occuring in foetal ovaries selectively eliminate oocytes with synaptic abnormalities. Subsequent studies by other workers on the mouse have shown that univalent frequency in cultured adult mouse oocytes is dependent on the preparative technique (Sugawara and Mikamo 1986). The type and size of univalent recorded at MI has also been demonstrated to be unrelated to non-disjunction as scored at MII (Sugawara and Mikamo 1983); cross-over frequency does not appear to decrease with maternal ageing in the mouse (Beermann et al. 1987).

Alternative approaches to the origin of non-disjunction studied at Edinburgh have examined the relationship between the bivalents and meiotic spindle apparatus. Eichenlaub-Ritter et al. (1988) have shown little evidence of MI spindle changes in aged CBA mouse oocytes. However, changes in the timing of oocyte development were observed, the germinal vesicle showing a delay in its breakdown and progression to MI. Subsequently, a more rapid passage than normal occurred in the MI to anaphase I stage. A similar effect has also been recorded by Hummler et al. (1987) for the rate of progression of hamster oocytes through meiosis.

Increased human usage of hazardous substances within the environment has stimulated investigations of the potential for mutagens to increase the levels of human aneuploidy. Analysis of the data relating to women exposed to therapeutic X-rays and the incidence of Down's syndrome (trisomy 21) has shown no clear-cut picture of involvement (Evans et al. 1986). Many mutagens have been shown increase the numbers of univalents seen at MI in spermatogenesis or oogenesis in the mouse (Hansmann 1983; Allen et al. 1986), whereas no relationship between univalency and transmission to the MII stage has been proven. Liang et al. (1986) investigated this problem in male mice treated with either 1 Gy whole body irradiation (Cobalt 60) or chemical mutagens. Having scored MII stages, they found no correlation between MI univalents and subsequent levels of non-disjunction. In the female, mouse oocytes showed no significantly increased levels of non-disjunction after low levels of X-irradiation (5 cGY), as measured at MII of meiosis. Only when irradiation levels reached 80 cGy were numbers of hypoploid MII oocytes significantly increased over controls (Hansmann et al. 1982).

It would seen likely that other factors involving hormonal control of the reproductive system and the time scheduling of meiosis must be investigated to come to terms with the problems of non-disjunction and aneuploidy in man.

5 Future Molecular Approaches to Meiosis

While molecular genetics has rapidly advanced into all fields of somatic cell genetics, with an ever-increasing wealth of knowledge on gene structure and function, it is only just beginning to make its impact felt on the mammalian meiotic system. This is partly due to the difficulty in obtaining experimental tissue and germ cells from either male or female in large enough quantities. No reliable tissue culture system as yet exists; the interaction of the male germ cells, both with themselves via intercellular bridges, and with the Sertoli cells, and that of the oocyte with the follicular cells, makes such an approach fraught with difficulties. The development of male germ cell sedimentation techniques has at least allowed the preparation of relatively homogenous cell suspensions. This has allowed the detailed biochemical analysis of the substages of meiosis, particularly by Hotta and Stern (1975). More recent molecular approaches to germ cell biochemistry have involved the extraction of specific RNA transcripts, such as the poly (A+) RNA transcribed during zygotene in both *Lilium* and mouse. Hotta et al. (1985) have suggested that such sequences play an important role in homologue pairing. The further use of mitotic cDNA libraries and cells at various stages of meiosis should then allow the identification of many genes which are specifically expressed during mammalian spermatogenesis. In this way Willison and Ashworth (1987) have described that both the protamine and PGK-2 genes are transcribed predominantly in post-meiotic haploid germ cells. Further, eight protooncogenes have been reported (Propst et al. 1988) as being actively transcribed in the germ cells of both male and female mice. Protooncogenes are implicated in tumour formation in adult mice and humans. Non-disjunction in tumors leads to widely fluctuating chromosome complements and alterations in growth control; could abnormal gene function in germ cells also promote meiotic non-disjunction? An alternative approach to the analysis of meiotic function would be to utilise transgenic mice. Peschon et al. (1987) using such technology have shown that transcription of the protamine 1 gene in post-meiotic round spermatids is dependent on a 2.4 kbp DNA sequence of which only 880 bp is 5' to the start of transcription. Transgenic mice may also aid the understanding of the process to genomic imprinting. It has become evident that the parental genetic contributions to the mammalian zygote are not equivalent in their eventual mode of action. A transgenic c-myc gene inherited from the father is only expressed in heart tissue, no expression occurring if inherited from the mother (Swain et al. 1987). Imprinting has been found to persist through DNA replication and cell division, and the process has the ability of switching the imprint from one sex to the other in alternating generations. Differential methylation of the genomes has been proposed to explain the phenomenon. Paternally transmitted genes so far implicated in the process show reduced levels of methylation. Questions as to the number of imprinted genes, and their relative contribution to genetic diseases such as Huntington's chorea, where the symptoms appear earlier in life if the gene is inherited from the father, remain to be answered.

Molecular techniques will also lead to a better interpretation of the structure and function of the meiotic chromosomes. Towards the understanding of synapsis, the composition of isolated synaptonemal complexes has been studied by several groups using either DNA probes (Li et al. 1983) or monoclonal antibodies (Moens et al. 1987; Heyting et al. 1988). The initial results show that major components of the SC are specific only to the meiotic prophase, and that proteins involved with the lateral elements are not involved in meiotic chromatin condensation. Molecular probes will also aid the investigation of meiotic chromosome structure at the first metaphase. In situ hybridization has shown (Mitchell et al. 1986) that the paramere region associated with the stretched centromeric heterochromatin of human bivalent No. 9 is composed of one family of repeated DNA sequences. The unusual nature of this region at meiosis may reflect a state of undermethylation. Further understanding of the structure of the meiotic chromatin should also come from the recent adaptation of the "nick translation" techniques developed to examine mitotic chromatin activity (Kerem et al. 1983). DNAse sensitivity is a reflection of chromatin conformation which may be related to potential gene transcription. In both the mouse (Rajcan-Separovic and Chandley 1987; Richler et al. 1987) and man (Chandley and McBeath 1987), sensitive sites, corresponding to early replicating mitotic bands, are evident along the sex bivalent. One such site may involve the so-called X chromosome inactivation center. While the human XY pairing segment also appears to be DNAse sensitive, there is some debate as to the nature of the pairing region in the mouse. The nick-translation technique should also allow investigation of the meiotic chromatin status in situations of subfertility caused by translocations. Air-dried pachytene cells from such mice have initially shown lower levels of nicking, possibly reflecting germ cell death (Erasmus B., personal communication 1988). Surface spreading combined with nick translation (Fig. 10) indicates that when X/autosome contact occurs in such

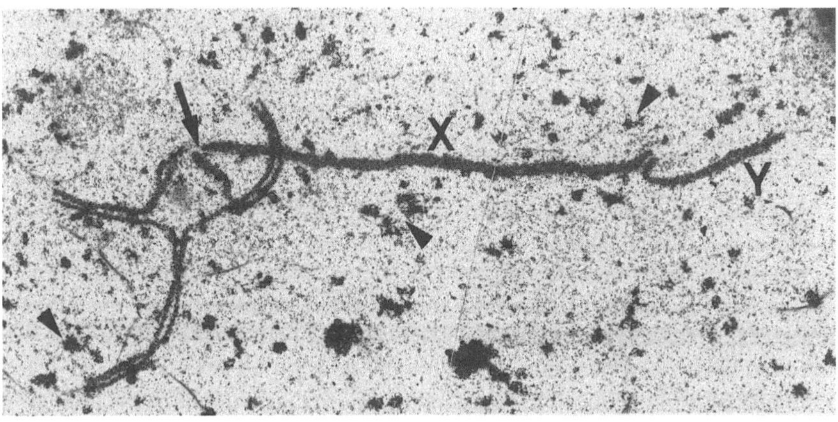

Fig. 10. Nick translation at the EM level in translocations causing male sterility. Mouse T(11;19) 42H quadrivalent showing contact with the XY bivalent *(arrow)*. Autoradiographic grains uniformly distributed in the chromatin around the quadrivalent and XY bivalent *(arrowheads* indicate a few of the grains present)

translocations, grain distribution is uniform, there being no apparent redistribution of nicking activity reflecting either X chromosome reactivation or autosomal suppression.

Finally, molecular probes are being utilized to refine the study of the parental origin of the extra chromosome in aneuploid individuals. Chromosomal polymorphisms originally permitted such identification in limited numbers of informative cases. The use of restriction-fragment length polymorphisms and polymorphic centromere probes now makes it possible to determine parental origin in virtually every case. Trisomy 21, where 80% of errors occur during the first maternal meiotic division (Hassold et al. 1984), can be contrasted with Klinefelter's syndrome. Here the additional X can now be seen to be inherited from either parent (maternal 47%:paternal 53%), the majority of extra paternal chromosomes deriving from the first meiotic division (Jacobs et al. 1988).

In the light of the complexity of the subject, and considering the limited number of people engaged in meiotic research, it is to be hoped that advances continue to be made. The basic, but as yet little understood, questions of mammalian meiotic pairing, recombination and segregation are of immense interest, as they influence the future of the organism from embryogenesis to fertility and disease susceptibility in adult life.

References

Allen JW, Liang JC, Carrano A, Preston JR (1986) Review of literature on chemical-induced aneuploidy in mammalian male germ cells. Mutat Res 167: 123–137

Allan JW, De Weese GK, Gibson JB, Poorman PA, Moses MJ (1987) Synaptonemal complex damage as a measure of chemical mutagen effects on mammalian germ cells. Mutat Res 190: 19–24

Ambros PF, Sumner AT (1987) Correlation of pachytene chromomeres and metaphase bands of human chromosomes, and distinctive properties of telomere regions. Cytogenet Cell Genet 44: 223–228

Andrle M, Fiedler W, Rett A, Ambros P, Schweizer D (1979) A case of trisomy 22 in *Pongo pygmaeus*. Cytogenet Cell Genet 24: 1–6

Ashley T (1984) A re-examination of the case for homology between the X and Y chromosomes of mouse and man. Hum Genet 67: 372–377

Ashley T, Russel LB, Cacheiro NLA (1983) Synaptonemal complex analysis of X-7 translocations in male mice: R2 and R6 quadrivalents. Chromosoma (Berl) 88: 171–177

Backer LC, Gibson JB, Moses MJ, Allen JW (1988) Synaptonemal complex damage in relation to meiotic chromosome aberrations after exposure of male mice to cyclophosphamide. Mutat Res 203: 317–330

Baker TG (1963) A quantitative and cytological study of germ cells in human ovaries. Proc R Soc Lond B Biol Sci 158: 417–433

Baker TG (1972) Oogenesis and ovulation. In: Austin CR, Short RV (eds) Reproduction in mammals. I. Germ cells and fertilization, Cambridge Univ Press, Lond, chap 2

Batanian J, Hultén MA (1987) Electron microscopic investigations of synaptonemal complexes in an infertile human male carrier of a percentric inversion inv(1)(p32q42). Regular loop formation but defective synapsis including a possible interchromosomal effect. Hum Genet 76: 81–89

Beechey CV (1973) X-Y chromosome dissociation and sterility in mouse. Cytogenet Cell Genet 12: 60–67

Beermann F, Bartels I, Franke U, Hansmann I (1987) Chromosome segregation at meiosis I in female T (2;4)1 Gö/+ mice: no evidence for a decreased crossover frequency with maternal age. Chromosoma (Berl) 95: 1–7

Berthelsen JG, Skakkebaek NE, Perbøll O, Neilson J (1981) Electron microscopical demonstration of the X and Y chromosomes in spermatocytes from human XYY males. In: Byskov AG, Peters H (eds) Development and function of reproductive organs. Int Congr Ser No 559. Elsevier/North Holland, p328

Blazak WF (1987) Incidence of aneuploidy in farm animals. In: Vig BK, Sandberg AA (eds) Aneuploidy Part A: incidence and etiology. Alan R Liss, Inc, NY, pp 103–116

Bond DJ, Chandley AC (1983) Aneuploidy. Oxford Univ Press, Oxford, NY, Toronto (Oxford monographs and medical genetics, No 11)

Bojko M (1983) Human meiosis VIII. Chromosome pairing and formation of the synaptonemal complex in oocytes. Carlsberg Res Commun 48: 457–483

Brewen JG, Preston RJ, Gengozian N (1975) Analysis of X-ray induced chromosomal translocations in human and marmoset spermatogonial stem cells. Nature (Lond) 253: 468–470

Brook JD (1983) X-chromosome segregation, maternal age and aneuploidy in the XO mouse. Genet Res Camb 41: 85–95

Brook JD, Chandley AC (1985) Testing of 3 chemical compounds for aneuploidy induction in the female mouse. Mutat Res 157: 215–220

Burgoyne PS (1982) Genetic homology and crossing-over in the X and Y chromosomes of mammals. Hum Genet 61: 85–90

Burgoyne PS, Baker TG (1984) Meiotic pairing and gametogenic failure. In: Evans CW, Dickinson HG (eds) Controlling events in meiosis. Company of Biologists, Cambrigde, p349

Burgoyne PS, Baker TG (1985) Perinatal oocyte loss in XO mice and its implications for the aetiology of gonadal dysgenesis in XO women. J Reprod Fertil 75: 633–645

Burgoyne PS, Biddle FG (1980) Spermatocyte loss in XYY mice. Cytogenet Cell Gent 28: 143–144

Carr DH, Haggar RA, Hart AG (1968) Germ cells in the ovaries of XO female infants. Am J Clin Pathol 49: 521–526

Cattanach BM (1962) XO mice. Genet Res 3: 487–490

Cattanach BM, Pollard CE, Hawkes SG (1971) Sex-reversed mice: XX and XO males. Cytogenetics (Basel) 10: 318–337

Cawood AH and Breckon G (1983) Synaptonemal complexes as indicators of induced structural change in chromosomes after irradiation of spermatogonia. Mutat Res 122: 149–154

Chandley AC (1971) Culture of mammalian oocytes. J Reprod Fertil Suppl 14: 1–6

Chandley AC (1975) Human meiotic studies. In: Emery AEH (ed) Modern Trends in Human Genetics. Butterworth, Lond, chap 2

Chandley AC (1979) The chromosomal basis of human infertility. Br Med Bull 35: 181–186

Chandley AC (1984) Infertility and chromosomal abnormality. In: Clarke JR (ed) Oxford reviews of reproductive biology, vol 6. Oxford Univ Press, Oxford, chap 1

Chandley AC (1986) A model for effective pairing and recombination at meiosis based on early replicating sites (R-bands) along chromosomes. Hum Genet 72: 50–57

Chandley AC, Fletcher J (1973) Centromere staining at meiosis in man. Humangenetik 18: 247–252

Chandley AC, McBeath S (1987) DNAse I hypersensitivity sites along the XY bivalent at meiosis in man include the XpYp pairing region. Cytogenet Cell Genet 44: 22–31

Chandley AC, Speed RM (1987) Cytological evidence that the Sxr fragment of XY, Sxr mice pairs homologously at meiotic prophase with the proximal testis-determining region. Chromosoma (Berl) 95: 345–349

Chandley AC, Fletscher J, Robinson JA (1976) Normal meiosis in two 47 XYY men. Hum Genet 33: 231–240

Chandley AC, Goetz P, Hargreave JB, Joseph AM, Speed RM (1984) On the nature and extent of XY pairing at meiotic prophase in man. Cytogenet Cell Genet 38: 241–247

Chandley AC, Speed RM, McBeath S, Hargreave TB (1986) A human 9;20 reciprocal translocation associated with male infertility. Cytogenet Cell Genet 41: 145–153

Chandley AC, McBeath S, Speed RM, Yorston L, Hargreave TB (1987) Pericentric inversion in human chromosome 1 and the risk for male sterility. J Med Genet 24: 325–334

Coerdt W, Rehder H, Gausmann I, Johannisson R, Gropp A (1985) Quantitative histology of human fetal testes in chromosome disease. Pediatr Pathol 3: 245–259

Cooke HJ, Brown WRA, Rappold GA (1985) Hypervariable telomeric sequences from the human sex chromosomes are pseudoautosomal. Nature (Lond) 317: 687–692

Counce SJ, Meyer GF (1973) Differentiation of the synaptonemal complex and the kinetochore in *Locusta* spermatocytes studied by whole-mount electron microscopy. Chromosoma (Berl) 44: 231–253

Curry CR, Magenis RE, Brown M, Lanman JT, Tsai J, O'Lague P, Goodfellow P, Mohandas T, Bergner EA, Shapiro LJ (1984) Inherited chondrodysplasia punctata due to a deletion of the terminal short arm of an X chromosome. N Engl J Med 311: 1010–1015

De Boer P (1986) Chromosomal causes for fertility reduction in mammals. In: Serres FJ de (ed) Chemical mutagens, Vol 10. Plenum Publishing Corporation, NY, pp 427–467

De Boer P, Branje HEB (1979) Association of the extra chromosome of tertiary trisomic male mice with the sex chromosome during first meiotic prophase and its significance for impairment of spermatogenesis. Chromosoma (Berl) 73: 369–379

Dellarco VL, Voytek PE, Hollaender A (1985) Aneuploidy: etiology and mechanisms. Plenum Press, NY

Eichenlaub-Ritter U, Chandley AC, Gosden RG (1988) The CBA mouse as a model for age-related aneuploidy in man: studies of oocyte maturation, spindle formation and chromosome alignment during meiosis. Chromosoma (Berl) 96: 220–226

Epstein CJ (1985) Mouse monosomies and trisomies as experimental systems for studying mammalian aneuploidy. Trends Genet 1: 129–134

Esposito MS (1984) Molecular mechanisms of recombination in *Saccharomyces cerevisiae:* testing mitotic and meiotic models by analysis of hypo-rec and hyper-rec mutations. In: Evans CW, Dickinson HG (eds) Controlling events in meiosis. Company of Biologistis Cambridge, p123

Evans EP, Breckon G, Ford CE (1964) An air-drying method for meiotic preparations from mammalian testes. Cytogenetics (Basel) 3: 289–294

Evans EP, Burtenshaw MD, Cattanach BM (1982) Meiotic crossing-over between the X and Y chromosomes of male mice carrying the sex reversing (Sxr) factor. Nature (Lond) 300: 443–445

Evans EP, Ford CF, Chaganti RSK, Blank CE, Hunter H (1970) XY spermatocytes in an XYY male. Lancet i: 719–720

Evans HJ (1967) The nucleolus, virus infection and trisomy in man. Nature (Lond) 214: 361–363

Evans HJ, Lyon MF, Czeizel A, Alberman ED, Ayme S, Baird PA, Beres J, Carothers AD, Chandley AC, Crouchley HC, Ferguson-Smith MA, Hassold TJ, Hook EB, Kallen B, Lindsten J, Mikkelsen M, Nevin NC, Ratcliffe SG, Tease C, Warburton D (1986) ICPEMC Meeting Report No 3, Is the incidence of Down syndrome increasing? Mutat Res 175: 263–266

Ferguson-Smith MA (1966) X-Y chromosomal interchange in the aetiology of true hermaphroditism and of XX Klinefelter's syndrome. Lancet 2: 475–476

Fletcher JM (1979) Light microscope analysis of meiotic prophase chromosomes by silver staining. Chromosoma (Berl) 72: 241–248

Ford CE, Hamerton JL (1956) The chromosomes of man. Nature (Lond) 178: 1020–1023

Forejt J (1974) Non-random association between a specific autosome and the male mouse: possible consequence of the homologous centromeres separation. Cytogenet Cell Genet 13: 369–383

Forejt J (1979) Meiotic studies of translocations causing male sterility in the mouse. II. Double heterozygotes for Robertsonian translocations. Cytogenet Cell Genet 23: 163–170

Forejt J (1981) Hybrid sterility gene located in the T/t-H-2 supergene on chromosome 17. In: Reisfeld RA, Ferrone S (eds) Current trends in histocompatability vol 1. Plenum Press, NY, p103

Forejt J (1982) X-Y involvement in male sterility caused by autosome translocations – a hypothesis. In: Crosignani PG, Rubin BL, Fraccaro M (eds) Genetic control of gamete production and function. Academic Press, Lond, NY, pp 135–151

Gabriel-Robez O, Ratomponirina C, Dutrillaux B, Carre-Pigeon F, Rumpler Y (1986a) Meiotic association between the autosomal quadrivalent of a reciprocal translocation in two infertile men, 46XY, t(19;22) and 46,XY,t(17;21). Cytogenet Cell Genet 43: 154–160

Gabriel-Robez O, Ratomponirina C, Rumpler Y, Le Marec B, Luciani M, Guichaoua MR (1986b) Synapsis and synaptic adjustment in an infertile human male heterozygous for a pericentric inversion in chromosome 1. Hum Genet 72: 148–152

German J (1968) Mongolism, delayed fertilizaton and human sexual behaviour. Nature (Lond) 217: 516–518

Gilliavod N, Leonard A (1973) Sensitivity of the mouse oocytes to the induction of translocations by ionizing radiations. Can J Genet Cytol 15: 363–366

Gillies CB (1973) Ultrastructural analysis of maize pachytene karyotypes by three dimensional reconstruction of the synaptonemal complexes. Chromosoma (Berl) 43: 145–176

Goodfellow P, Banting G, Sheer D, Ropers HH, Caine A, Ferguson-Smith MA, Povey S, Voss R (1983) Genetic evidence that a Y-linked gene in man is homologous to a gene on the X chromosome. Nature (Lond) 302: 346–349

Goodfellow P, Darling S, Wolfe J (1985) The human Y chromosome. J Med Genet 22: 329–344

Gosden CM, Wright MO, Paterson WG, Grant KA (1976) Clinical details, cytogenetic studies, and cellular physiology of a 69,XXX fetus, with comments on the biological effect of triploidy in man. J Med Genet 13: 371–380

Gropp A, Winking H (1981) Robertsonian translocations: cytology, meiosis, segregation patterns and biological consequences of heterozygosity. Zool Soc Lond Symp 47: 141–181

Gropp A, Winking H, Redi C (1982) Consequences of Robertsonian heterozygosity: segregational impairment of fertility versus male-limited sterility. In: Crosignani PG, Rubin BL (eds) Genetic control of gamete production and function. Academic Press Lond, NY, p115

Guichaoua MR, Delafontaine D, Taurelle R, Taillemite JL, Morazzani MR, Luciani JM (1985) Loop formation and synaptic adjustment in a human male heterozygous for two pericentric inversions. Chromosoma (Berl) 93: 313–320

Hackstein JHP (1987) Spermatogenesis in *Drosophila.* In: Henning W (ed) Spermatogenesis genetic aspects. Springer, Berlin Heidelberg New York Tokyo pp 63–116

Hale DW (1986) Heterosynapsis and suppression of chiasmata within heterozygous pericentric inversions of Sitka deer mouse. Chromosoma (Berl) 94: 425–432

Handel MA (1987) Genetic control of spermatogenesis in mice. In: Hennig W (ed) Spermatogenesis genetic aspects. Springer, Berlin Heidelberg New York Tokyo pp 1–62

Hansmann I (1983) Factors and mechanisms involved in non-disjunction and X-chromosome loss. In: Sandberg A (ed) The cytogenetics of the mammalian X-chromosome. Liss, NY, pp 131–170

Hansmann I, Jenderny J, Probeck HD (1982) Non-disjunction and chromosome breakage in mouse oocytes after various X-ray doses. Hum Genet 61: 190–192

Hassold TJ, Chiu D, Yamane JA (1984) Parental origin of autosomal trisomies. Ann Hum Genet 48: 129–144

Heller CG, Wootton P, Rowley MJ, Lalli MF, Bruscan DR (1975) Action of radiation upon human spermatogenesis. Excerpta Med Int Congr Ser VI Pan Am Congr Endocrinol No 112, p408

Henderson SA, Edwards RG (1968) Chiasma frequency and maternal age in mammals. Nature (Lond) 218: 22–28

Heyting C, Dettmers RJ, Dietrich AJJ, Redeker EJW, Vink ACG (1988) Two major components of synaptonemal complex are specific for meiotic prophase. Chromosoma (Berl) 96: 325–332

Hotta Y, Stern H (1975) Zygotene and pachytene – labelled sequences in the meiotic organization of chromosomes. In: Peacock WJ, Brock RD (eds) The Eukaryote Chromosome. Aust Acad Sci Canberra, pp 283–300

Hotta Y, Tabata S, Stubbs L, Stern H (1985) Meiosis-specific transcripts of a DNA component replicated during chromosome pairing: homology across the phylogenetic spectrum. Cell 40: 785–793

Hultén M (1974) Chiasma distribution in the normal human male. Hereditas 76: 55–78

Hultén M, Pearson PL (1971) Fluorescent evidence for spermatocytes with two Y chromosomes in an XYY male. Ann Hum Genet 34: 273–276

Hultén MA, Saadallah N, Batanian J (1987) Meiotic chromosome pairing in the human male: experience from surface spread synaptonemal complexes. Chromosomes Today 9: 218–229

Hummler E, Theuring F, Hansmann I (1987) Meiotic non-disjunction in oocytes from aged Djungarian hamsters correlates with an alteration in meiosis rate but not in univalent formation. Hum Genet 76: 357–364

Hungerford DA (1971) Chromosome structure and function in man. I. Pachytene mapping in the male, improved methods and general discussion of initial results. Cytogenetics (Basel) 10: 23–32

Jacobs PA, Hassold TJ, Whittington E, Butler G, Collyer S, Keston M, Lee M (1988) Klinefelter's syndrome: an analysis of the origin of the additional sex chromosome using molecular probes. Ann Hum Genet 52: 93–109

Jacobs PA, Szulman AE, Funkhouser J, Matsuura JS, Wilson CC (1982) Human triploidy: relationship between parental origin of the additional haploid complement and development of partial hydatidiform mole. Ann Hum Genet 46: 223–231

Jagiello GM (1978) Reproduction in Down syndrome. In De la Cruz FF, Gerald PS (eds) Trisomy 21 (Down syndrome). Univ Park Press, Baltimore, p151

Jagiello GM, Kamicki J, Ryan RJ (1968) Superovulation with pituitary gonadotrophins. Method for obtaining metaphase figures in human ova. Lancet 1: 178–180

Jagiello GM, Lin JS, Ducayen MB (1975) SO_2 and its metabolite: effects on mammalian egg chromosomes. Environ Res 9: 84–93

Jagiello GM, Ducayen MB, Miller WA, Lin JS, Fang JS (1973) A cytogenetic analysis of oocytes from *Macaca mulatta* and *Nemestrina* matured in vitro. Humangenetik 18: 117–122

Johannisson R, Gropp A, Winking H, Coerdt W, Rehder H, Schwinger E (1983) Down's syndrome in the male. Reproductive pathology and meiotic studies. Hum Genet 63: 132–138

Johannisson R, Löhrs U, Passarge E (1988) Pachytene analysis in males heterozygous for a familial translocation (9;12;13)(q22;q22;q32) ascertained through a child with partial trisomy 9. Cytogenet Cell Genet 47: 160–166

Johannisson R, Löhrs U, Wolff HH, Schwinger E (1987) Two different XY-quadrivalent associations and impairment of fertility in man. Cytogenet Cell Genet 45: 222–230

Joseph AM, Chandley AC (1984) Morpholigical sequence of XY pairing in the Norway rat *Rattus norvegicus*. Chromosoma (Berl) 89: 381–386

Kalikinskaya EI, Kolomiets OL, Shevchenko VA, Bogdanov Yu F (1986) Chromosome aberrations in F_1 from irradiated male mice studied by their synaptonemal complexes. Mutat Res 174: 59–65

Kerem B, Goitein R, Richler C, Marcus M, Cedar H (1983) In situ nick translation distinguishes between active and inactive chromosomes. Nature (Lond) 304: 88–90

Kimball RF (1987) The development of ideas about the effect of DNA repair on the induction of gene mutations and chromosomal aberrations by radiation and by chemicals. Mutat Res 186: 1–34

Kjessler B (1974) Chromosomal constitution and male reproductive failure. In: Mancini RE, Martini L (eds) Male fertility and sterility. Academic Press, Lond, NY, p231

Koller PC, Darlington CD (1934) The genetical and mechanical properties of the sex chromosomes. I. *Rattus norvegicus*. J Genet 29: 159–173

Koulischer L, Schoysman R (1975) Étude des chromosome mitotiques et meiotiques chez les hommes infertiles. J Génét Hum 23: 58–70

Kremer H, Hennig W, Dijkhof R (1986) Chromatin organisation in the male germ line of *Drosophila hydei*. Chromosoma (Berl) 94: 147–161

Leenhouts HP, Chadwick KH (1981) An analytical approach to the induction of translocations in the spermatogonia of mouse. Mutat Res 82: 305–321

Li S, Meistrich ML, Brock WA, Hsu TC, Kuo MT (1983) Isolation and preliminary characterization of the synaptoneml complex from rat pachytene spermatocytes. Exp Cell Res 144: 63–72

Liang JC, Sherron DA, Johnston D (1986) Lack of correlation between mutagen-induced chromosomal univalency and aneuploidy in mouse spermatocytes. Mutat Res 163: 285–297
Lifschytz E, Hareven D (1977) Gene expression and the control of spermatid morphogenesis in *Drosophila melanogaster.* Dev Biol 58: 276–294
Lifschytz E, Lindsley PL (1972) The role of X-chromosome inactivation during spermatogenesis. Proc Natl Acad Sci USA 69: 182–186
Lindsley DL, Sandler L (1977) The genetic analysis of meiosis in female *Drosophila melanogaster.* Phil Trans R Soc Lond B, 277: 295–312
Luciani JM, Stahl A (1971) Etude des states de debut de la meiose chez l'ovocyte foetal humain. Bull Assoc Anat 15: 445–458
Luciani JM, Guichaoua MR, Mattei A, Morazzani MR (1984) Pachytene analysis of a man with a 13q;14q translocation and infertility. Behaviour of the trivalent and non-random association with the sex vesicle. Cytogenet Cell Genet 38: 14–22
Luiciani JM, Guichaoua MR, Delafontaine D, North MO, Gabriel-Robez O, Rumpler Y (1987) Pachytene analysis in a 17;21 reciprocal translocation carrier: role of the acrocentric chromosomes in male sterility. Hum Genet 77: 246–250
Lyon MF, Cox BD (1975) The induction by X-rays of chromosome aberrations in male guinea-pigs, rabbits and golden hamsters. III. Dose-response relationship after single doses of X-rays to spermatogonia. Mutat Res 29: 407–422
Lyon MF, Hawker SG (1973) Reproducive life span in irradiated and unirradiated XO mice. Genet Res 21: 185–194
Lyon MF, Meredith R (1966) Autosomal translocations causing male sterility and viable aneuploidy in the mouse. Cytogenetics 5: 335–354
Lyon MF, Cox BD, Marston JH (1976) Dose-response data for X-ray induced translocations in spermatogonia of rhesus monkeys. Mutat Res 35: 429–436
Madan K, Hompes PGA, Schoemaker J, Ford CE (1981) X-autosome translocation with a breakpoint in Xq22 in a fertile woman and her 47XXX infertile daughter. Hum Genet 59: 290–296
Mahadevaiah S, Mittwoch U (1986) Synaptonemal complex analysis in spermatocytes and oocytes of tertiary trisomic Ts(5^{12})31H mice with male sterility. Cytogenet Cell Genet 41: 169–176
Martin RH, Balkan W, Burns K, Rademaker AW, Lin CC, Rudd NL (1983) The chromosome constitution of 1000 human spermatozoa. Hum Genet 63: 305–309
Martin RH, Hildebrand K, Yamamoto J, Rademaker A, Barnes M, Douglas G, Arthur K, Ringrose T, Brown IS (1986) An increased frequency of human sperm chromosomal abnormalities after radiotherapy. Mutat Res 174: 219–225
Matsuda Y, Tobari I, Yamagiwa J, Utsugi T, Kitazume M, Nakai S (1984) γ-ray induced reciprocal translocations in spermatogonia of the crab-eating monkey (*Macaca fasicularis*). Mutat Res 129: 373–380
Mayer B, Schellander K, Auer H, Tesarik E, Schleger W, Sasshofer K, Glawischnig E (1987) Offspring of a trisomic cow. Cytogenet Cell Genet 44: 229–230
Maynard-Smith J (1978) The evolution of sex. Cambridge Univ Press, Lond NY Melbourne
McIlree ME, Selby-Tulloch W, Newsam JE (1966) Studies on human meiotic chromosomes from testicular tissue. Lancet i: 679–682
Mendel GJ (1865) Versuche über Pflanzen-Hybriden. Verh Naturforsch Ver Brünn 4: 3–47
Miklos GLC (1974) Sex chromosome pairing and male fertility. Cytogenet Cell Genet 13: 558–577
Mitchell AR, Ambros P, McBeath LS, Chandley AC (1986) Molecular hybridisation to meiotic chromosomes in man reveals sequence arrangement on the No. 9 chromosome and provides clues to the nature of "parameres". Cytogenet Cell Genet 41: 89–95
Mittwoch U, Mahadevaiah S, Olive MB (1981) Retardation of ovarian growth in male-sterile mice carrying an autosomal translocation. J Med Genet 18: 414–417
Moens PB (1973) Quantitative electron microscopy of chromosome organisation at meiotic prophase. Cold Spring Harbor Symp Quant Biol 38: 99–107
Moens PB, Heyting C, Dietrich AJJ, Van Raamsdonk W, Chen Q (1987) Synaptonemal complex antigen location and conservation. J Cell Biol 105: 93–103

Monesi V (1972) Spermatogenesis and the spermatozoa. In: Austin CR, Short RV (eds) Reproduction of mammals. I. Germ cells and fertilization. Cambridge Univ Press, Lond, chap 3

Moses MJ (1977a) Microspreading and the synaptonemal complex in cytogenetic studies. In: De la Chapelle A, Sorsa M (eds) Chromosomes today, vol 6. Elsevier/North Holland Biomed Press, Amsterdam p71

Moses MJ (1977b) Synaptonemal complex karyotyping in spermatocytes of the Chinese hamster (*Cricetulus griseus*). II. Morphology of the XY pair in spread preparations. Chromosoma (Berl) 60: 127–137

Moses MJ (1981) Meiosis, synaptonemal complex, and cytogenetic analysis. In: Jagiello G, Vogel HJ (eds) Bioregulators of reproduction. Academic Press, Lond, NY, p187

Moses MJ, Poorman PA (1981) Synaptonemal complex analysis of mouse chromosome rearrangements. II. Synaptic adjustment in tandem duplications. Chromosoma (Berl) 81: 519–535

Moses MJ, Poorman PA (1984) Synapsis, synaptic adjustment and DNA synthesis in mouse oocytes. Chromosomes Today 8: 90–103

Moses MJ, Poorman PA, Roderick TH, Davisson MT (1982) Synaptonemal complex analysis of mouse chromosome rearrangements. IV. Synapsis and synaptic adjustment in two paracentric inversions. Chromosoma (Berl) 84: 457–474

Moses MJ, Poorman PA, Russell LB, Cacheiro NL, Roderick TH, Davisson MT (1978) Synaptic adjustment: Two pairing phases in meiosis? J Cell Biol 79: 123a

Muller HJ (1928) The measurement of gene mutation rate in *Drosophila*, its high variability and its dependence upon temperature. Genetics 13: 279–357

Oakberg EF (1957) Gamma-ray sensitivity of spermatogonia of the mouse. J Exp Zool 134: 343–356

Oakberg EF (1960) Gamma ray sensitivity of oocytes of young mice. Anat Rec 137: 385–386

Oakberg EF, Di Minno RL (1960) X-ray sensitivity of primary spermatocytes of the mouse. Intern J Radiat Biol 2: 196–209

Ohno S, Kaplan W, Kinosita R (1959) Do XY- and O- sperm occur in *Mus musculus?* Exp Cell Res 18: 382–384

Ohno S, Makino S, Kaplan WD, Kinosita R (1961) Female germ cells of man. Exp Cell Res 24: 106–110

Pearson PL, Ellis JD, Evans HJ (1970) A gross reduction in chiasma formation during meiotic prophase and a defective DNA repair mechanism associated with a case of human male infertility. Cytogenetics (Basel) 9: 460–467

Penrose LS (1933) The relative effects of paternal and maternal age in mongolism. J Genet 27: 219–224

Peschon JJ, Behringer RR, Brinster RL, Palmiter RD (1987) Spermatid-specific expression of protamine 1 in transgenic mice. Proc Natl Acad Sci USA 84: 5316–5319

Peters H, Byskov AG, Grimsted J (1981) The development of the ovary during childhood in health and disease. In: Coutts JRT (ed) Functional morphology of the human ovary. pp 26–34

Polani PE (1972) Centromere localization at meiosis and the position of chiasma in the male and female mouse. Chromosoma (Berl) 36: 343–374

Polani P, Dewhurst J Sir, Fergusson I, Kelberman J (1982) Meiotic chromosomes in a female with primary trisomic Down's syndrome. Hum Genet 62: 277–279

Power MM (1987) Equine half sibs with an unbalanced X;15 translocation or trisomy 28. Cytogenet Cell Genet 45: 163–168

Propst F, Rosenberg MP, Vande Woude GF (1988) Proto-oncogene expression in germ cell development. Trends in Genet 4: 183–187

Quack B, Speed RM, Luciani JM, Noel B, Guichaoua M, Chandley AC (1988) Meiotic analysis in two human reciprocal X-autosome translocations. Cytogenet Cell Genet 48: 43–47

Rajcan Separovic ER, Chandley AC (1987) Lack of evidence that the XqYq pairing tips at meiosis in the mouse show hypersensitivity to DNAse I. Chromosoma (Berl) 95: 290–294

Ratomponirina C, Andrianivo J, Rumpler Y (1982) Spermatogenesis in several intra- and interspecific hybrids of the lemur (*Lemur*). J Reprod Fertil 66: 717–721

Rees H (1961) Genotypic control of chromosome form and behaviour. Bot Rev 27: 288–318
Richler C, Uliel E, Kerem B, Wahrman J (1987) Regions of active chromatin conformation in active male meiotic sex chromosomes of the mouse. Chromosoma (Berl) 95: 167–170
Riley R, Flavell RB (1977) A first view of the meiotic process. Phil Trans R Soc Lond B 277: 191–199
Röhrborn G (1970) The activity of alkylating agents. I. Sensitive mutable stages in spermatogenesis and oogenesis. In: Vogel F, Röhrborn G (eds) Chemical mutagenesis in mammals and man. Springer, Berlin Heidelberg New York, chap 19
Rosenmann A, Wahrman J, Richler C, Voss R, Persitz A, Goldman B (1985) Meiotic association between the XY chromosomes and unpaired autosomal elements as a cause of human male sterility. Cytogenet Cell Genet 39: 19–29
Rouyer F, Simmler M C, Johnsson C, Vergnaud G, Cooke HJ, Weissenbach J (1986) A gradient of sex linkage in the pseudoautosomal region of the human sex chromosome. Nature (Lond) 319: 291–295
Rudak E, Jacobs PA, Yanagimachi R (1978) Direct analysis of the chromosome constitution of human spermatozoa. Nature (Lond) 274: 911–913
Russell LB (1976) Numerical sex-chromosome anomalies in mammals: their spontaneous occurrence and use in mutagenesis studies. In: Hollaender A (ed) Chemical mutagens, vol 4. Plenum, NY, p55
Russell P, Altschuler G (1975) The ovarian dysgenesis of trisomy 18. Pathology 7: 149–155
Saadallah N, Hultén MA (1985) A complex three breakpoint translocation involving chromosomes 2,4 and 9 identified by meiotic investigations of a human male ascertained for subfertility. Hum Genet 71:312–320
Saadallah N, Hultén MA (1986) EM investigations of surface spread synaptonemal complexes in a human male carrier of a pericentric inversion inv(13)(p12q14). The role of heterosynapsis for spermatocyte survival. Ann Hum Genet 50: 369–383
Sadasiviah RS, Kasha KJ (1971) Meiosis in haploid barley – an interpretation of non-homologous chromosome associations. Chromosoma (Berl) 35: 247–263
Sarto GE, Therman E, Patau K (1973) X inactivation in man: a woman with t(Xq−;12q+). Am J Hum Genet 25: 262–270
Schleiermacher E (1970) The Activity of Alkylating Agents. II. Historical and cytogenetic findings in spermatogenesis. In: Vogel F, Röhrborn G (eds) Chemical mutagenesis in mammals and man. Springer, Berlin Heidelberg New York, p167
Searle AG (1982) The genetics of sterility in the mouse. In: Crosignani PG, Rubin BL, Fraccaro M (eds) Genetic control of gamete production and function. Academic Press, Lond NY, p93
Shobha Devi P, Polasa H (1987) Evaluation of the anti-inflammatory drug indomethacin for its genotoxicity in mice. Mutat Res 188: 343–347
Simmler M-C, Rouyer F, Vetgnand G, Nyström-Lahti M, Ngo KY, De la Chapelle A, Weissenbach J (1985) Pseudoautosomal DNA sequences in the pairing region of the human sex chromosomes. Nature Lond 317: 692–697
Skakkebeak NE, Hultén M, Jacobson P, Mikkelsen M (1973) Quantification of human seminiferous epithelium. II. Histological studies in eight XYY men. J Reprod Fertil 32: 391–401
Smith A, Fraser IS, Elliot G (1979) An infertile male with balanced Y;19 translocation. Review of Y; autosome translocations. Ann Genet 22: 189–194
Solari AJ (1970) The spatial relationship of the X and Y chromosomes during meiotic prophase in mouse spermatocytes. Chromosoma (Berl) 29: 217–236
Solari AJ (1980) Synaptonemal complexes and associated structures in microspread human spermatocytes. Chromosoma (Berl) 81: 315–337
Solari AJ, Tres L (1970) The three-dimensional reconstruction of the XY chromosome pair in human spermatocytes. J Cell Biol 45: 43–53
Speed RM (1984) Meiotic configurations in female trisomy 21 fetuses. Hum Genet 66: 176–180
Speed RM (1985) The prophase stages in human foetal oocytes studied by light and electron microscope. Hum Genet 69: 69–75
Speed RM (1986a) Prophase pairing in a mosaic 18p-;iso18q human female fetus studied by surface spreading. Hum Genet 72: 256–259

Speed RM (1986b) Oocyte development in XO foetuses of man and mouse: the possible role of heterologous X-chromosome pairing in germ cell survival. Chromosoma (Berl) 94: 115–124

Speed RM (1988) The possible role of meiotic pairing anomalies in the atresia of human fetal oocytes. Hum Genet 78: 260–266

Speed RM, Chandley AC (1983) Meiosis in the foetal mouse ovary. II. Oocyte development and age-related aneuploidy. Does a production line exist? Chromosoma (Berl) 88: 184–189

Stahl A, Hartung M, Devictor M, Berge-Lefranc JL (1984) The association of the nucleolus and the short arm of acrocentric chromosomes with the XY pair in human spermatocytes: its possible role in facilitating sex-chromosome acrocentric translocations. Hum Genet 68: 173–180

Stearn PE, Droulard KE, Stahhar FH (1960) Studies bearing on fertility of male and female mongoloids. Am J Ment Defic 65: 37–41

Sugawara S, Mikamo K (1983) Absence of correlation between univalent formation and meiotic non-disjunction in aged female Chinese hamsters. Cytogenet Cell Genet 35: 34–40

Sugawara S, Mikamo K (1986) Maternal ageing and non-disjunction: a comparative study of two chromosomal techniques on the formation of univalent in first meiotic metaphase oocytes of the mouse. Chromosoma (Berl) 93: 321–325

Sumner AT, Speed RM (1987) Immunochemical labelling of the kinetochore of human synaptonemal complexes, and the extent of pairing of the X and Y chromosomes. Chromosoma (Berl) 95: 359–365

Swain JL, Stewart TA, Leder P (1987) Parental legacy determines methylation and expression of an autosomal transgene: A molecular mechanism for parental imprinting. Cell 50: 719–727

Tateno H, Mikamo K (1984) Neonatal oocyte development and selective oocyte-killing by X-rays in the Chinese hamster, *(Cricetulus griseus)*. Int J Radiat Biol 45: 139–149

Templado C, Vidal F, Navarro J, Marina S, Egozcue J (1984) Meiotic studies and synaptonemal complex analysis in two infertile males with a 13/14 balanced translocation. Hum Genet 67: 162–165

Thomson E, Fletcher J, Chandley AC, Kůcerová M (1979) Meiotic and radiation studies in four oligochiasmatic men. J Med Genet 16: 270–277

Thorsland TW, Paulsen CA (1972) Proc Natl Symp Natural and man-made radiation in space. NASA Document TMX-2440, pp 229–232

Tiepolo L, Zuffardi O, Fraccaro M, Giarola A (1981) Chromosome abnormalities and male infertility. In : Frajese G, Hafez ES, Conti C, Fabbrini A (eds). Oligozoospermia: recent progress in andrology. Raven Press, NY, p 233

Tres LL (1975) Nucleolar RNA synthesis of meiotic prophase spermatocytes in the human testis. Chromosoma (Berl) 53: 141–151

Uebele-Kallhardt B, Knörr K (1971) Meiotische Chromosomen der Frau. Humangenetik 12: 182–187

Van Buul PPW (1984) X-ray induced translocations in marmoset *(Callithrix jacchus)* stem cell spermatogonia. Mutat Res 129: 231–234

Vig BK, Sandberg AA (1987) Aneuploidy. Part A: Incidence and etiology. Part B: Induction and model systems. Alan R Liss Inc, NY

Von Wettstein D, Rasmussen SW, Holm PB (1984) The synaptonemal complex in genetic segregation. Ann Rev Genet 18: 331–413

Wallace BMN, Hultén MA (1983) Triple chromosome synapsis in oocytes from a human fetus with trisomy 21. Ann Hum Genet 47: 271–276

Wallace BMN, Hultén MA (1985) Meiotic chromosome pairing in the normal human female. Ann Hum Genet 49: 215–226

Westergaard M, Von Wettstein D (1970) Studies on the mechanism of crossing-over. IV. The molecular organisation of the synaptonemal complex in *Neottiella* (Cooke) Saccardo (Ascomycetes). CR Trav Lab Carlsberg 37: 239–268

Westergaard M, Von Wettstein D (1972) The synaptonemal complex. Ann Rev Genet 6: 71–110

Willison K, Ashworth A (1987) Mammalian spermatogenic gene expression. Trends in Genet 3: 351–355

Yuncken C (1968) Meiosis in the human female. Cytogenetics (Basel) 7: 234–238

Mutagen-Mutation Equilibria in Evolution

H. NÖTHEL[1]

Contents

1 Introduction

Mutations are double-faced like Janus. They are most often detrimental to their carriers – they also represent the ultimate source of genetic variability and, hence, of evolution. This antagonism requires a delicate balance. It may be achieved by one or more of three different strategies:

1. Mutations do not entirely occur at random in the genome. This is evident for specific types of somatic mutations in mammals (Baltimore 1981; Golub 1987; Meyer et al. 1986) and for transposon-induction of mutations in the germ line (Engels and Preston 1984; O'Hare and Rubin 1983; Rubin et al. 1982; see also Campbell 1983, and Syvanen 1984).

2. Mutation rates are adjusted to evolutionary requirements. This has since long been discussed controversially (Benado et al. 1976; Biémont et al. 1987; Chao et al. 1983; Cox and Gibson 1974; Dobzhansky et al. 1952; Gillespie 1981; Holsinger and Feldman 1983; Ives 1950; Nöthel 1983, 1987; Sturtevant 1937) but seems more likely in view of the surprising mixture of error-proof and error-prone (i.e. mutagenic) mechanisms of DNA repair (Kimball 1987).

3. In diploids, the genetic load due to recessive variants is not a mutational load that has to be avoided in order to gain optimal population fitness, but mainly is a tolerable "segregational noise" produced by selectively favoured heterozygotes according to the "balance-hypothesis".

[1] Institut für Allgemeine Genetik, Freie Universität Berlin, Arnimallee 7, D-1000 Berlin-33

How populations actually maintain a balance between mutational benefits and genetic death is not only of interest for understanding basic evolutionary processes, but has important practical implications in view of the continuously elevated concentrations of environmental mutagens. The question has been approached by theoretical models (Gillespie 1981; Holsinger and Feldman 1983) and by experimental studies (Benado et al. 1976; Chao et al. 1983; Cox and Gibson 1974; Dobzhansky et al. 1952). However, experimental evidence is poor, because any valid study requires (1) several hundreds of generations with increased mutagenic activity, (2) a Mendelian population as a test system to get a situation comparable to man and most "higher" animals and plants, and (3) a variety of appropriate techniques in population genetics and mutation research. The only experimental organism that fits these requirements is *Drosophila.*

In long-term studies of experimental populations of *D. melanogaster* I have tried to analyze the evolutionary response to an elevated environmental input of mutations. I will present the phylogeny and the general adaptations of populations subjected to drastically and permanently increased X-ray exposures for up to 750 generations (*RÖ*-populations). Recent data will illustrate the continual adjustment, by a variety of mechanisms, of the delicate balance between genetic load and genetic variability. Genetically controlled mechanisms of mutagen resistance will be shown to be a main source of adaptation, fine-tuning the adjustment of mutation rates according to mutagen input. The result of these findings: the hypothesis of mutagen-mutation equilibria in populations. Evidence will be given on achievement of part of this resistance by transposons, and the general role of movable DNA in mutagen-mutation equilibria in evolution is discussed. Finally, I will consider contributions of some sterilizing effects of mutagens to such an equilibrium: if only one local population is exposed to a higher input of mutations, sterility prevents migration over the entire gene pool.

2 Phylogeny of Irradiated *RÖ-Populations of Drosophila melanogaster* and Their Adaptations to X-Rays

Phylogeny and irradiation histories of the *RÖ*-populations are illustrated in Fig. 1. Common origin of all populations is the "Berlin-wild" stock of *Drosophila melanogaster.* It was collected from nature in Berlin prior to 1930 and has been kept since then as a laboratory wild-type stock. It was split up into several sublines (+ or *+60* or *+K)* and these were used to initiate the irradiated populations. These are labeled *RÖI, RÖII* etc. which refers to the first, second, etc. population exposed to X-*(Röntgen)*-rays. Subpopulations are designated by suffixes, i.e., $RÖI_4$ indicates a subpopulation of *RÖI* that is exposed to 4.4 kR (instead of the original 2.2 kR), and $RÖI_{40}$ is a subpopulation of $RÖI_4$ without further exposures (1 kR = 0.258 C/kg). *RÖX* stands for a subpopulation of $RÖI_{48}$ that is exposed to X-rays of increasing, maximum tolerable levels.

All *RÖ*-populations became adapted to their new environment "increased ionizing radiation". Adaptation is simple in *RÖIII* and *RÖIV.* Figure 2 demon-

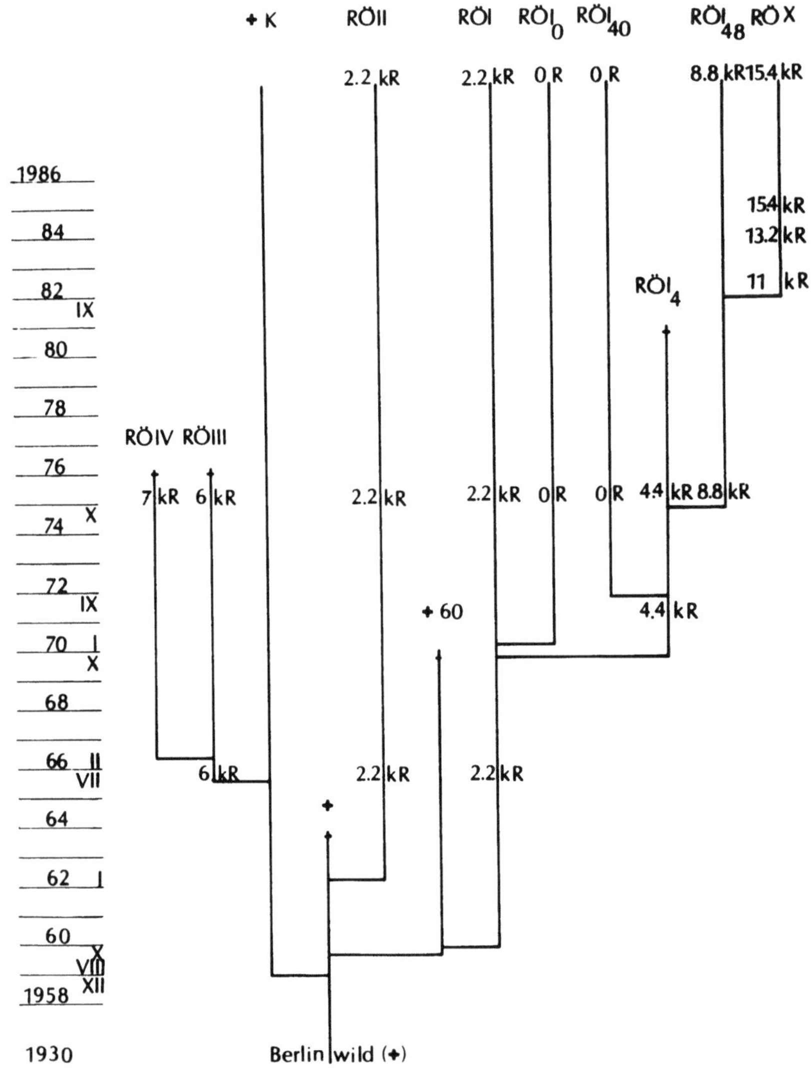

Fig. 1. Phylogeny of *Berlin-wild* sublines, of the X-irradiated populations *RÖI* to *RÖIV* and of several subpopulations of *RÖI*. The two *left columns* give year (*Arabic numbers*) and month (*Roman numbers*) when the populations or subpopulations were started or terminated. 26 generations were reared per year in *Berlin wild, RÖI* (including its subpopulations) and *RÖII*, and 17 generations in *RÖIII* and *RÖIV*. Actual irradiation levels to which adult flies of a population were exposed in every generation are incluced in the phylogenetic tree (note that exposures have been recalculated following improvements of dosimetry and may diverge up to 10% from nominal levels given in earlier publications). Germ cell stages exposed in *RÖI, RÖII*, and *RÖI4* were stages 6–14 oocytes and mature sperm and spermatids, in $RÖI_{48}$ and *RÖX* only stages 6–14 oocytes were irradiated, and in *RÖIII* and *RÖIV* oogonia and spermatogonia were treated. Population size was 500 pairs (in 20 250 ml bottles with progeny thoroughly mixed in every generation), and this was increased to 1000 pairs in $RÖI_4$, $RÖI_{48}$, and *RÖX*

strates that, in both populations with irradiations of gonial germ cell stages, reproductive capacity is initially decreased but almost immediately begins to recover in spite of continuing irradiations. Analysis of this adaptation proved it to be due to selection against sort of a detrimental factor homozygous in the basic population Berlin wild: the genetic factor *av* (*Ablege-Verhinderung*, Nöthel 1969) prevents oviposition after X-irradiation. It is induced at exposures above 2 kR and is 100% effective at 4 kR, i.e., at 4 kR and above, females carrying this factor in effective dose are permanently sterilized. The factor is recessive, located on chromosome 4, and of incomplete penetrance due to genetic modifiers that can easily be selected. It has no apparent effect in unirradiated flies. After irradiation it prevents oviposition, not egg maturation. As a consequence, mature eggs accumulate, ovaries enlarge, and the female dies within a few days because it cannot get rid of the egg masses (Nöthel 1969). In *RÖIII* and *RÖIV*, with repro-

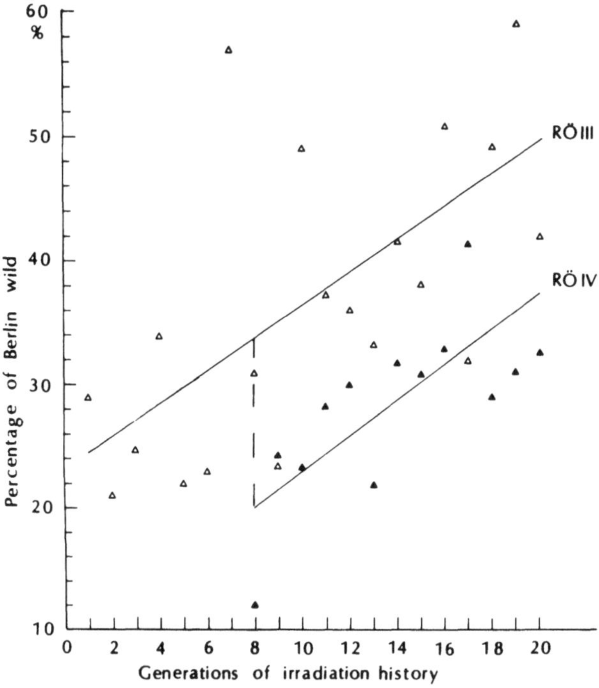

Fig. 2. Adaption of viability in *RÖIII* (△) and *RÖIV* (▲) wihin 20 generations of irradiation history. Viability was measured by oviposition of surviving females on post-irradiation days 22–26, and this is immediately after the actual reproductive period of the population. Viability is expressed as "mean No. of eggs per female × No. of surviving females/No. of surviving flies". The correction includes survival of males; this is appropriate since with the exposures used, male survival is unaffected and, hence, reflects variations between generations in culture conditions. The resulting viability was corrected by the corresponding value obtained from $+K$ as long time mean and is given in percent of this *Berlin wild* viability. Note that at initiation of a subpopulation with higher per generation exposure (*RÖIII* from $+K$, *RÖIV* from *RÖIII*) viability decreased markedly but that it immediately started to recover. According to analysis of regression (regression lines calculated are given in the figure), regression is linear and significant (α = 0.004 in *RÖIII*, α = 0.008 in *RÖIV*) in the generations considered

ductive periods on post-irradiation days 18–21, selection against this factor was quite successful. It changed the modifier system and thus increased dramatically post-irradiation oviposition and female survival. This simple type of selection response has to be expected whenever there is a genetic factor that is detrimental to its carriers under irradiation conditions, and that is either heterozygous, or can be modified by other genes.

Other *RÖ*-populations likewise became adapted to X-rays (Fig. 3). Severe reductions in progeny numbers, induced in any population when initiated at an increased exposure level, are followed by subsequent readjustments of population numbers to the limits given by the carrying capacity of the system. This adaptation

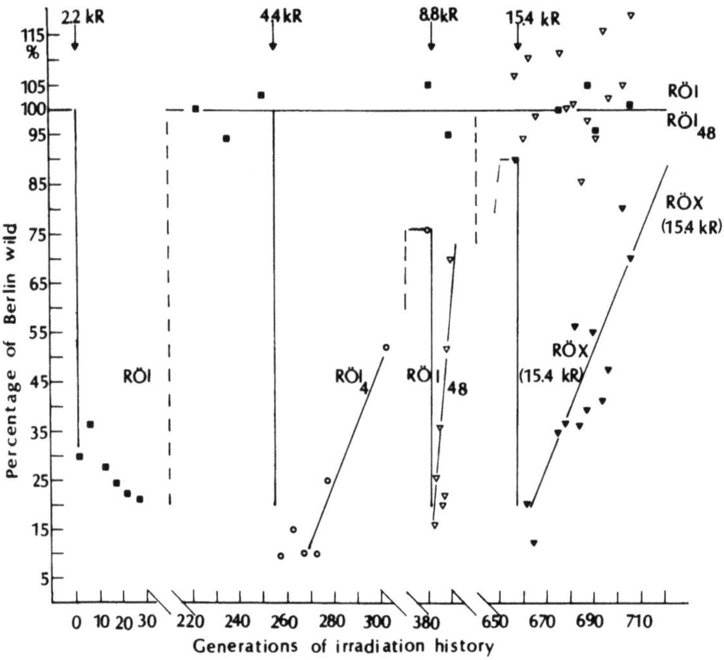

Fig. 3. Adaptation of viability in *RÖI* (■) and its irradiated subpopulations *RÖI*$_4$ (○), *RÖI*$_{48}$ (▽) and *RÖX* (▼) in course of irradiation history. Progeny numbers obtained, in any generation, from the irradiated *RÖ* are given in percent of those contemporaneously obtained from unirradiated *Berlin wild*. *Berlin wild* was reared like *RÖI* except for the irradiations. Progeny numbers were calculated per bottle (always 20), because in *Berlin wild* progeny per bottle is limited by the carrying capacity of a bottle and not by the number of parents per bottle. Changes in progeny numbers during generations that were not tested are indicated by *broken lines*. At initiation of a subpopulation with higher per generation exposure (*RÖI* from *Berlin wild*, *RÖI*$_4$ from *RÖI*, *RÖI*$_{48}$ from *RÖI*$_4$, *RÖX* from *RÖI*$_{48}$ and finally *RÖX* from *RÖX* at a lower exposure), progeny numbers dropped abruptly due to dominant lethals and, due to accumulation of genetic load, the decrease continued over several generations. But progeny numbers finally came back to *Berlin wild* level in *RÖI* to *RÖI*$_{48}$ in spite of continued irradiations. In *RÖX*, a continuous selection response due to the steady increase in irradiation level is, for example, indicated by the recovery after increasing the per generation exposure from 13.2 to 15.4 kR. It is, according to regression analysis, linear and significant ($\alpha = 0.0002$), whereas in *RÖI*$_{48}$ there is no apparent alteration in viability anymore within the same interval of generations ($\alpha = 0.5474$)

is due to complex interactions. They are outlined in Fig. 4 with respect to *RÖI₄*. This figure highlights what was stated above on the "janus head" of mutations. In principle, irradiations produce mutations, and these exert considerable selection pressures either directly, via dominant lethals, or indirectly, via accumulation of recessive genetic load. A response to this pressure is possible by the induced genetic variability of all sorts. This selection response increases general viability and buffers recessive lethals (in a way still not understood) against becoming homozygous, and it enhances relative genetic radioresistance, so that mutagenic effects of the radiation input are considerably reduced.

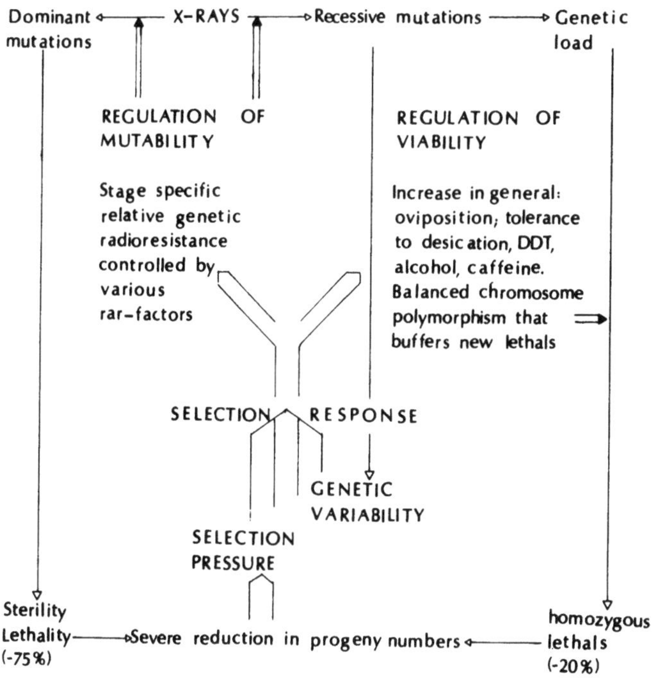

Fig. 4. Pathways of adaptation to high X-ray exposures in RÖI and its irradiated subpopulations. X-rays induce mutations. Dominant mutations predominantly produce sterility (by killing of germ cell stages) and lethality (by damages in germ cells transmitted to progeny). At initiation of a population (*RÖI*, *RÖI₄* etc), the yield of dominant mutations is at least 75%. Recessive mutations are most often lethal and contribute to the genetic load of a population. Any of the major autosomes carries one to several lethals with rates of allelism being about 12%. Therefore, at least 20% of the zygotes will die because of homozygosity of recessive lethals. The resulting amount of total lethality means a severe reduction in progeny numbers, and hence, a strong selection pressure. On the other hand, mutations of all sorts steadily increase genetic variability of the gene pool. Thereby, the raw material is provided for the selection pressure to shape on. The selection response is manifold. Viability has been regulated so that reproduction is increased and metabolism in general is improved (as measured by a variety of parameters). Interestingly, in some cases (third chromosomes of *RÖI₄*), a balanced inversion polymorphism "buffers" recessive lethals and prevents them from becoming homozygous. In another selection response, mutability has been down regulated: stage specific relative genetic resistance, controlled by different genetic factors, reduces considerably the mutagenic action of X-ray exposures

3 Adaptive Resistance of *RÖ*-Populations to the Induction of Genetic Radiation Damage and Genetic Factors Controlling It

Interest is focused on this relative resistance to the mutagenic effects of X-rays. Induction of various types of mutation, i.e., dominant lethals, sex-linked recessive lethals, numerical and structural aberrations, were measured after exposure of immature oocytes (stage 7, according to King 1970). It was carefully ascertained that no shifts occurred in treated germ cell stages sampled from the various populations. This is especially important since these stages may differ considerably in mutagen sensitivity (Sankaranarayanan and Sobels 1976). Moreover, all comparisons between populations and genotypes are based on dose-response regressions. These are linear with respect to recessive lethals; with respect to other types of mutation they are made straight lines by probit transformations (Fig. 5). Differences between populations are characterized by dose-reduction factors (DFRs), and these are quotients between "dose required to induce a given mutation rate in a less sensitive genotype" and "dose required to induce the same effect in the control genotype of normal sensitivity". DRFs are verified in common dose-response regressions, established by dividing actual exposures of less sensitive genotypes by their respective DRF (Fig. 6). The DRFs indicate fine-tuned evolutionary adjustments of radiosensitivities of the populations to the X-ray levels they are exposed to (Table 1). Sensitivity to the induction of genetic damage in immature oocytes decreases step by step from the control population (+*K*), to (1) populations irradiated at 2.2 kR per generation (*RÖI, RÖII*); (2) the population exposed to 4.4 kR per generation ($RÖI_4$); (3) the population exposed to 8.8 kR per generation ($RÖI_{48}$); (4) the population exposed to 11 and more kR per generation (*RÖX*). At each level, sensitivity is constant (or oscillates around a constant mean) over hundreds of generations. This mean apparently characterizes the optimal mutation rate under the given environmental conditions, i.e., irradiation levels.

Adaptation to optimal mutation rates is achieved by distinct and different genetic factors of relative radioresistance in the various RÖ-populations. These

Fig. 5. a, b. Radiosensitivities of *Berlin wild* and various *RÖ*-populations: percentage of dominant lethals induced in immature oocytes (predominantly stage 7) of +*K* (+), *RÖII* (◆), *RÖI* (■), $RÖI_4$, $RÖI_{48}$ (▽), and *RÖX* (▼). Dominant lethality was measured by egg-to-adult survival. Frequency of induced lethality (*i*) was calculated by using the formula $i = (u - c)/(1 - c)$, where *1-u* is egg-to-adult survival after irradiation and *1-c* is survival in the control group. Data were obtained in 1986, and > 1000 eggs were tested per experimental point. Sigmoid dose response regressions were made straight lines by probit transformation. **a** Regression lines given are from earlier data (Nöthel 1987). Experimental points are from 1986 data. They do fit the lines. *RÖX* recently gained a significant resistance compared to $RÖI_{48}$, and this can be described by a DRF of 1.22 (yielding a DRF of 4.6 compared to +*K*). **b** DRF values estimated are verified by means of a dose response regression common for all populations. It is based on effective doses (x scale) obtained by dividing actual exposures by the respective DRF. DRF values were 1.0 (+*K*), 1.7 (*RÖI, RÖII*), 3.8 ($RÖI_{48}$), and 4.6 (*RÖX*)

radiation resistance factors (*rar*) have been isolated, located within the genome, and characterized with respect to their effects (as measured by means of DRFs) on different types of mutation, to their stage specificity (life cycle and germ cell stages), and to some miscellaneous observations (Table 2). As expected from selection responses, *rar* factors are quite different in different populations kept

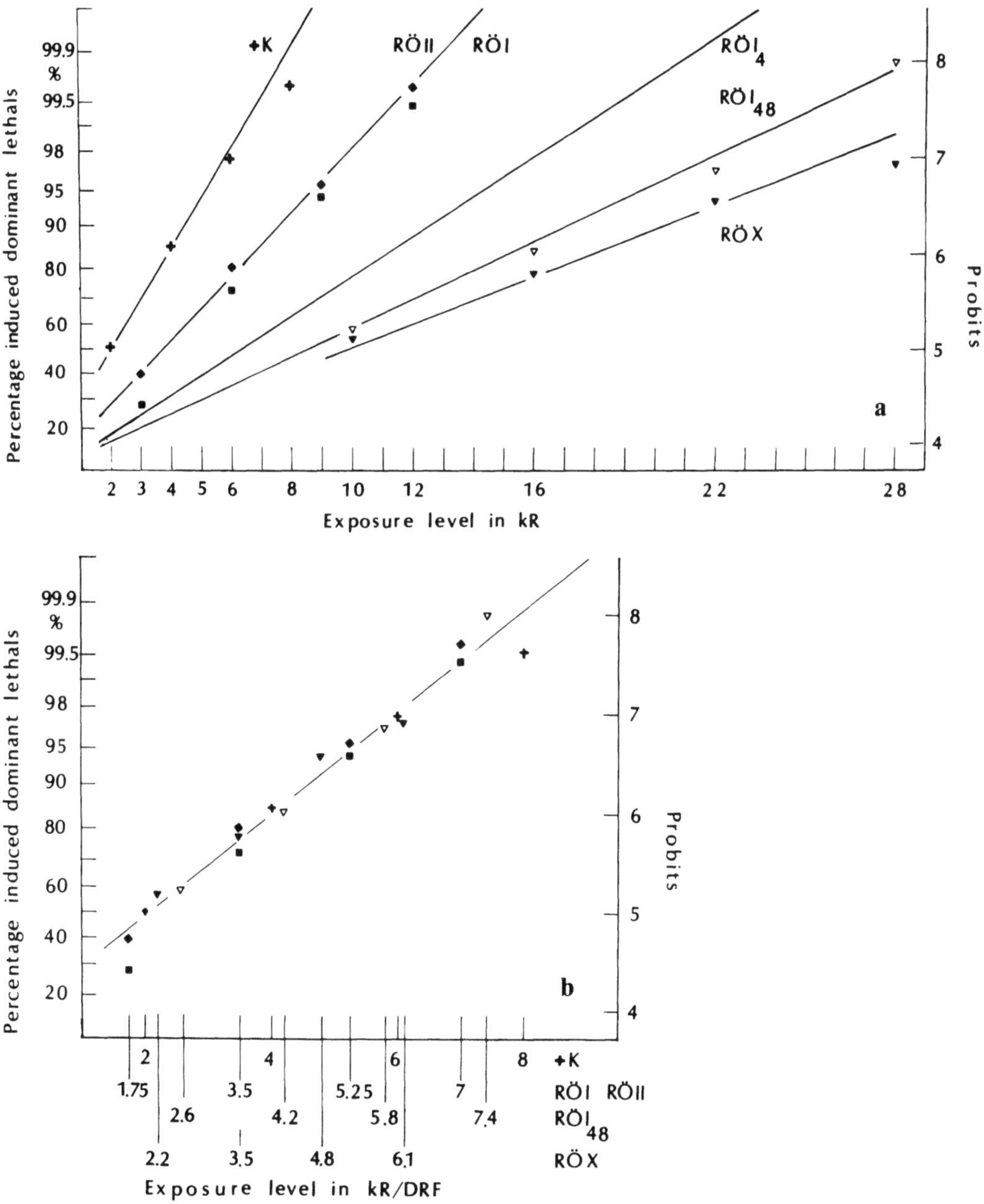

under the same irradiation conditions (compare *RÖI* and *RÖII*), but their overall effects are similar. Likewise, different factors evolved in subpopulations of *RÖI* under increasing irradiation levels. One point has to be stressed: *rar* factors were always newly induced in the respective *RÖ*-populations; they are recessive or semidominant and never surfaced in an ancestor kept at its per generation exposure level. These observations lead to the conclusion that mutagen-mutation equilibria are maintained, but not via polymorphism and heterosis.

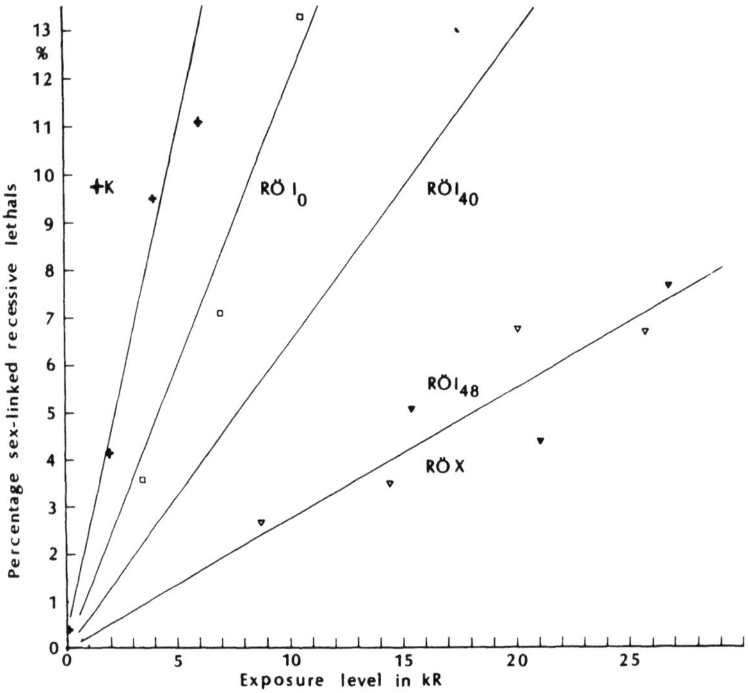

Fig. 6. Relative radioresistance of $RÖI_{48}$ and *RÖX*: sex-linked recessive lethals in percent (%) of X-chromosomes tested (N) after X-irradiation of immature oocytes (stage 7) of +*K* (+), $RÖI_0$ (□), $RÖI_{48}$ (▽), and *RÖX* (▼). Data were obtained in 1986, and only orthodox lethals were scored with the usual techniques [Nöthel and Weber (1976); Y-suppressed lethals were omitted]. For comparisons, earlier data on +*K*, $RÖI_0$, and $RÖI_{40}$ (Nöthel and Weber 1976) are represented by linear dose response regressions. Recent data do not deviate significantly from these lines, and this is indicated by experimental points given for +*K* and $RÖI_0$ (N is about 300 per point). $RÖI_{48}$ and *RÖX* do not differ in radiation response, and a common dose response regression does not significantly diverge from linearity (N is about 1000 per experimental point). The DRF of $RÖI_{48}$ and *RÖX* compared to +*K* is 7.5.

Females of $RÖI_{48}$ and *RÖX* are continuously irradiated. They may be heterozygous for pre-existing, sex-linked recessive lethals. Frequency of these lethals is two times the mutation rate and is reduced by one-half in any generation without exposure. Since one generation was unirradiated prior to all tests reported here, rate of pre-existing lethals equals the mutation rate induced by the per generation exposure, i.e., 8.8 kR in $RÖI_{48}$ and 15.4 kR in *RÖX*. Since only irradiated oocytes about stage 7 were used in the populations (all mature stages 14 are killed, all stages 6 are prevented from further maturation at the doses applied), these exposures were simply added to the irradiation levels administered in the tests, to achieve correction for pre-existing lethals

Table 1. Radioresistance of *RÖ*-populations relative to *Berlin wild:* DRF values with respect to the induction of dominant lethals in immature oocytes in course of population history[a]

Test period (year)	0.0 kR *Berlin wild* → DRF	2.2 kR *RÖII* DRF	2.2 kR *RÖI* → DRF	0.0 kR $RÖI_0$ DRF	4.4 kR $RÖI_4$ → DRF	0.0 kR $RÖI_{40}$ DRF	8.8 kR $RÖI_{48}$ → DRF	15.4 kR *RÖX* DRF
1968	1.0	1.64	1.71					
1969	1.0		1.8					
1971	1.0		1.8		1.8			
1972	1.0		1.68	1.72	2.5			
1972	1.0		1.82		2.78	2.78		
1973	1.0	1.72	1.72					
1974	1.0					3.23		
1975	1.0		1.72	1.72	2.72	2.72		
1976	1.0					3.2	3.1	
1980	1.0	1.7	1.7	1.7	2.8	2.0	2.78	
1983	1.0		1.7	1.7		1.9	3.8	3.8
1986	1.0	1.7	1.7	1.7		1.8	3.8	4.6

[a] *Top rows* give per generation exposure levels (in kR) of females of the various populations; *arrows* indicate descendance of these populations.

4 The Role of Transposons in Relative Radioresistance of *RÖ*-Populations and in Maintaining Mutagen-Mutation Equilibria

A hint for explaining the equilibrium in genetic terms comes from *rar-3*. This factor arose early in the history of $RÖI_4$ and quickly became homozygous; this homozygosity required in several cases the movement of *rar-3* between inversions, which usually prevent recombination. Moreover, *rar-3* disappeared from $RÖI_{40}$ after some 200 generations without irradiations and from several homozygous and unirradiated lines (where *rar-3* was flanked by visible markers, and their persistence indicates that *rar-3* was not lost due to stock-contamination), and it was at least drastically reduced in frequency in the continuously irradiated population $RÖI_{48}$ with the upcoming of *rar-4* (Nöthel 1987). These observations are hard to explain with conventional genetics, but are understandable in view of a transposon-like nature of *rar-3*. This interpretation is supported by the following:

In *Drosophila melanogaster*, P-elements are a family of transposons (Bingham et al. 1982). They are present in P-strains and are activated in the germ line of first-generation hybrids with paternally transmitted P-factors and cytoplasm maternally inherited from lines without P (M-cytotype) (Kidwell et al. 1977a,b; Rio et al. 1986). Activation is indicated by "hybrid dysgenesis", a syndrome of the hybrids characterized by sterility in part of males and females, by male recombi-

Table 2. Genetic factors controlling relative radioresistance in *RÖ*-populations[a]

Population	Factor	Dominance	Location genetic	cytologic[b]	DRF values in stage-7 oocytes Lethality	X-loss	SLRL[c]
RÖ I	Total				1.72	1.72	1.73
	rar-1	Semidom.	1-?	?	1.31	1.0	1.31
	rar-2	Semidom.	2-50	?	1.31	1.72	1.31
RÖI$_4$	Total				2.72	2.72	3.23
	rar-1						
	rar-2						
	rar-3	Recessive	3-50.1	86E2-4; 87C6-7	1.58	1.58	1.87
RÖ I$_{48}$	Total				3.83	3.07	7.5
	rar-1						
	rar-2						
	(rar-3)						
	rar-4	Recessive	1-14.8	5C2; 5D5-6	1.87	1.87	3.8
RÖ II	Total				1.72	1.72	1.53
	rar-5	Semidom.	3-?		1.52	1.4	1.6
	?	Additional	polygenes?		1.13	1.23	1.0

[a] The *rar* factors act independently in the various populations, as revealed by multiplication of DRF values. An exception are *rar-3* – *rar-4* interactions that reduce total effects almost to *rar-4* level. In *RÖI*$_{48}$ (and *RÖX*), rar-3 disappeared or was inactivated.
Stage specificity: *rar-1* and *rar-2* are only effective in oocytes of prophase I after recombination and prior to a mature stage-14 oocyte; *rar-3* and *rar-4* are active throughout oogenesis except mature stages and only in adults.
Additional information: *rar-1* is polymorphic in various stocks, it is Y-suppressed and can be inhibited by caffeine; *rar-2* reduces association of heterologous, chiasmatic chromosomes in the chromocentre and thus decreases preconditions for heterologous interchanges; *rar-1,* and *rar-3* reduce the mutagenic effects of EMS, whereas *rar-4* does apparently not affect MMS-mutagenesis.
[b] Located by deficiency mapping. Boundaries of deletions that uncover *rar* are given.
[c] *Lethality* refers to induction of dominant lethals, *X-loss* to induction of X-chromosome losses, and *SLRL* to sex-linked recessive lethals.

nation (MR, referring to occurrence of usually absent intrachromosomal recombination in males), by mutations induced in both sexes (point mutations, structural and numerical aberrations), etc. (Fitzpatrick and Sved 1986; Kidwell et al. 1977a,b; Mackay 1986). Mutations can arise by insertion of a P-element into a coding sequence (O-Hare and Rubin 1983; Rubin et al. 1982) or by unprecise excision (Voelker et al. 1984). Both insertions and excisions are not restricted to P, but activation of intact P-elements, i.e., activation of P-transposase (Engels 1984; Rio et al. 1986), likewise triggers movements of other transposons and insertional sequences (IS) in the genome (Rubin et al. 1982). Therefore, if *rar-3* is a transposon, or if its effects are caused by insertion of movable DNA, *rar-3* should be a hot spot for induction of mutations and MR by P-activation. Since *Berlin wild K,* all *RÖ*-populations, and all *rar-3*-lines are of M-cytotype, i.e., without P-elements, P-activation can easily be achieved by mating *RÖ*- or *rar-3*-

Table 3. Induction of male recombination (MR) and recessive lethals by P-activation in *rar-3*/*Harwich* and *Normal*/*Harwich* hybrid males

		Hybrid males without P-activity from crosses *Harwich*-♀♀ × *M*-♂♂		– Hybrid males with P-activity from crosses *M*-♀♀ × *Harwich*-♂♂				
rar-3 genotype:		*Normal*/*rar-3*		*Normal*/*normal*			*Normal*/*rar-3*	
Effects tested		Number N of chromosomes tested	Effects in % of N	N	%	Lit. %	N	%
MR *ru – ca*[a] (total chrom 3)				1468	2.04	2.19		
cu – kar							2798	0.29
cu – sr				1468	0.07		4466	0.36
sr – e				1468	0.00		4466	0.00
Sex-linked recessive lethals[b]		623	0.23	3191	2.04	2.75	873	2.06
Third chrom.rec.lethals	*Normal* chrom.						2081	0.05
within *DF(3R)T-32*[c]	*rar-3* chrom.	1800	0.00				2385	0.67

[a] MR was measured for the total third chromosome using *ru cu ca*, a chromosome with the markers (map position according to recombination frequencies in females given in parentheses) of the chromosome was obtained using the same markers. In addition, MR was tested for the of the chromosome was obtained using the same markers. In addition, MR was tested for the region *cu* (50.0) – *kar* (51.7), because *rar-3* (50.1) is located between these markers.

[b] Sex-linked recessive lethals induced in hybrid males were scored using the Basc technique (hybrid males being the male parents of the test), all presumed lethals were retested and only lethals *sensu stricto* were included in the data.

[c] Tests were performed as follows: In G0 (G0 to G4 refer to subsequent generations), *Harwich* males were mated to females carrying lethal free third chromosomes with *rar-3* and flanking markers (*cu sr e*). P-mutagenesis occurs in the germ line of G1-hybrids. G1 hybdrid males were mated to females carrying appropriate markers on a balancer third chromosome. G2 female progeny were scored for third chromosomes being either of *rar-3* origin (identified by marker phenotypes) or from *Harwich* genotype (normal phenotype). They were then mated individually to males carrying *Df(3R)T-32.* In G3, any individual progeny were scored for absence of the class heterozygous for the deletion and the *rar-3*- or normal (*Harwich*)-chromosome, respectively. Absence of this class indicates P-induction of a recessive lethal in a sperm of a G1-hybrid. Note that MR is likewise detected from the same progeny by change in marker combinations. These MRs are included in the upper part of the table. In any case (lethal or MR), appropriate retests were done and only lethals *sensu stricto* are considered here.

Sources: Kidwell et al. (1977a, b); Markers used are described by Lindsley and Grell (1968).

females with males of the *Harwich* wild-type stock (best P-line known) and testing the hybrids for MR and mutagenesis.

MR is simply measured in hybrids from M × P matings by backcrosses to M-females and scoring of progeny. P is always *Harwich,* and M carries third chromosome markers, either with or without *rar-3* between them. Details and results are given in Table 3. Results clearly indicate a hot spot for MR within the *rar-3* region of *rar-3* chromosomes.

Mutations are measured by induction of recessive lethals either in the X-chromosome or within part of the third chromosome that is deleted in *Df(3R) T-32*; breakpoints of this deficiency are the cytological sites 86E2-4 and 87C6-7 (*rar-3* is located within the boundaries of this deletion, see Table 2). Details and results are given in Table 3. There is a hot spot for mutation induction in the *rar-3* region of *rar-3* chromosomes: lethal-induction within the region of *Df(3R)T-32* is significantly higher in *rar-3* compared to the normal (*Harwich*) chromosome of the same G1 hybrid males. It is, however, not increased in the X-chromosomes of the same hybrids. Hence, the expected 50-fold mutation rate in the X-chromosome (1000 vital loci) compared to the region deleted in *Df(3R)T-32* (about 20 loci) was observed in the normal chromosome but not in *rar-3* where the ratio was only 3–4.

Knowledge of the cytological site occupied by *rar-3*, and of its nature as movable DNA, opens pathways for molecular analysis of this "regulator of mutability" via in situ hybridization with known *Drosophila* mobile genetic elements and subsequent isolation. Until then, it remains an open speculation whether *rar-3* is a structural gene for DNA-repair using a transposable element as vehicle, or whether *rar-3* is a regulator that normally represses repair activity in specific stages of oogenesis and that is inactivated by insertion of a movable element.

In any case, evolutionary adjustment to higher mutagen input via reduced mutagen sensitivity is at least in part mediated by transposons. This indicates a major contribution of movable DNA elements to evolutionary relevant genetic variability, to adaptive responses, and to evolution in general (Campbell 1983; Charlesworth and Charlesworth 1983; Ginzburg et al. 1984; Syvanen 1984). It is in favour of transposons as evolutionary factor that their mobility is tightly regulated. In *Drosophila*, for example, copy number is an intrinsic regulator on the population level (Biémont 1986; Biémont and Aouar 1987; Syvanen 1984), and differential splicing of introns from the transposase transcript (Rio et al. 1986) makes the difference between germ line and somatic mutagenicity of P-factors. Transpositions may either occur in general bursts (Gerasimova et al. 1984, 1985) or in element specific explosions (Biémont et al. 1987). Increase in genetic variability by transpositions improves adaptive selection responses (Belyaeva et al. 1982; Chao et al. 1983), and such increase is induced in populations of low genetic variability with low copy numbers of movable DNA (Biémont et al. 1987). Hence, there is ample evidence that movable DNA plays a crucial role in controlling mutagen-mutation equilibria in evolution.

5 ENU-Treated *Drosophila* Populations and the Role of Sterility in Mutagen-Mutation Equilibria

X-rays induce all types of mutations and more or less enhance the natural mutation spectrum. Induction of genetic variability and genetic load is therefore accompanied by induction of chromosome aberrations that lead to sterility/dominant lethality. Thus, mutation induction exerts an immediate selection pressure. Selective adaptation to this pressure may therefore be expected to result in reduced mutagen sensitivity. It is an interesting question, whether or not a similar response is found with respect to mutagens that yield only minor rates of chromosome aberrations, compared to point mutations, and therefore have a mere indirect effect on selective forces via accumulation of genetic load. To the best of my knowledge, one of the strongest mutagens of this kind is the monofunctional, direct alkylating chemical mutagen N-Nitroso-N-Ethylurea (ENU). ENU induces genetic effects predominantly by ethylations of O^6 and N^7 in guanine. In DNA, alkylations of N^7 frequently lead to apurinic sites and subsequently to strand-breaks and aberrations, whereas O^6-alkylations result in O^6-methylguanin-thymin base pairing, frequently followed by point mutations (Doerjer et al. 1984; Loeb and Preston 1986; Williams and Shaw 1987). Both types of DNA damage require rather specific mechanisms of repair (Foster and Davis 1987; Kimball 1987; Radman and Wagner 1986; Yarosh 1985). With respect to ENU, the relation between O^6- and N^7-alkylations is relatively high and, consequently, point mutations are frequent relative to aberrations (and sterility/dominant lethals) (Russell et al. 1982; Vogel and Natarajan 1979; Vogel et al. 1985). In *Drosophila sperm* we found 25% sex-linked recessive lethals at ENU concentrations that do not yield any 2; 3 translocation or any measurable sterility/dominant lethality.

For the past 2 years (46 generations), we have maintained several *Drosophila* populations that were treated with ENU in every generation. ENU was administered *per os* to either larvae or adults; with either stage, three different populations were reared, each treated with another concentration of ENU (including shame treated controls). Induction of mutations was as expected with respect to sex-linked recessive lethals, accumulation of autosomal lethals, allozyme variants, and aberrations. However, viability of ENU populations decreased extremely rapidly, much more than could be expected from this load of mutations. Çürüksu has shown in her thesis (in preparation) that this is in part due to an accumulation of sterility in males and, to a lesser degree, in females. This sterility is not due to chromosome aberrations; it is transmitted, but not according to simple modes of inheritance. In any initial generation (as in any mutation experiment), sterility is induced only in low frequency, but it increases rapidly within a few generations under ENU treatment and declines after cessation of treatment. Accumulation of sterility eventually reaches a constant and population specific induction-selection equilibrium, but some populations had to be kept at reduced exposures to prevent extinction. On the other hand, at least in the ENU populations with larval treatment, there is a highly significant reduction in sensitivity to the ENU induction of sex-linked recessive lethals in spermatogonia: comparisons with the shame treated population yield DRFs (obtained from linear dose response regressions) of 1.3 and 1.8 in the populations treated at lower and higher concentrations per generation, respectively (Çürüksu, in preparation). This is puzzling: why should the selective response to the low selection pressure by induction of point mutations be more pronounced than to the very strong pressure by accumulation of sterility to $>70\%$? The only plausible answer is that of a causal relation between induction of sterility and induction of mutation. There may be a biochemical pathway that reduces ENU mutagenesis by processes that enhance sterility. A more probable explanation may be that some special polygenic mutations give rise to sterility; therefore, selection against mutability means selection against induction of sterility. (Ironically enough, with this interpretation, the anticipated differences in X-ray and ENU mutation spectra disappear.) Polygenic mutations are apparently much more frequent than ordinary mutations (Mukai 1964, 1979; Mukai et al. 1972), even if closer knowledge is missing (Ramel 1983). In the ENU populations, some independent hints on highly induced and accumulated rates of polygenic mutations come from heritable morphological "markers" like wing malformations. On the other hand, tests according to Mukai's protocol were negative, but this probably is because sterility is not recorded with that scheme, and because the polygenic mutations causing sterility may represent a special sector of the whole spectrum of polygenic mutation (Nöthel and Çürüksu, unpublished).

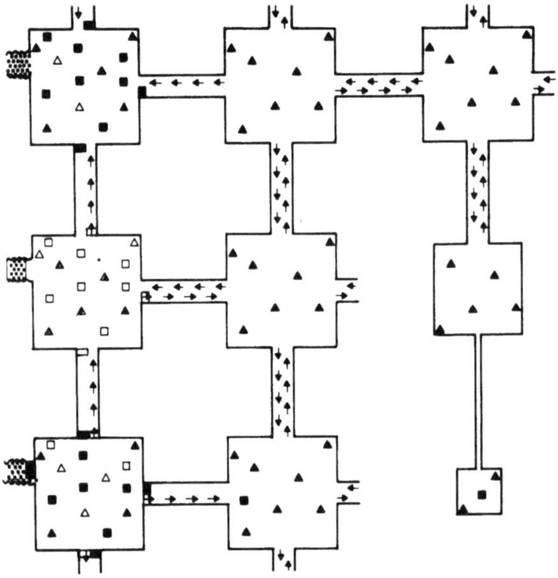

Fig. 7. Simplified model for maintenance of mutagen-mutation equilibra within the network of local populations that constitute a species. *Major squares* symbolize local populations of a species. Channels between them indicate migration and *arrows* direction of gene flow. *Dots* within *wavy lines* represent increases in environmental mutagens. *Triangles* symbolize normal genetic variability and *squares* indicate variants due to increases in mutagen input or in intrinsinc mutation rates, respectively. *Closed symbols* stand for an actual situation, *open ones* for a former variability that disappeared due to sterility or resistance, and *hatched signs* for repopulation by migration from neighbouring local populations. *Bars* indicate interruptions either of mutagenic effects (within *wavy lines*) or of migrational gene flow (within channels), and *open bars* refer to preadaptive situations.
Bottom right population is isolated and small and has low genetic variability. In this population, bursts of mutations occur especially by transposition. *Top left* population is at an early stage of increased exposure to mutagens. Accumulation of sterility results in unidirectional isolation that allows immigation but not emigration and prevents newly induced variants to spread over the entire population. With continuing mutagen-input the local population will either become extinguished or resistant. As a consequence of extinction, the deserted niche may later be repopulated from neighbouring populations as is illustrated in the *middle left square*. Evolution of adaptive resistance is represented in the *bottom left* population. It is accompanied by a switch in semipermeability of isolation: immigrants cannot reproduce under increased mutagen input but emigrants may spread over the entire genepool thus increasing evolution of the species

The most interesting aspects in the accumulations of sterility in populations under high mutagen input are evolutionary ones. On first sight, this decrease in viability is what is expected from the negative face of mutation in evolution. But to me, another scenario is much more attractive (Fig. 7): in terms of mutagen-mutation equilibria, mutagen resistance and mutagen-induced sterility may well be two pathways to reduce an environmental input of mutations into the gene pool of a population. In general, any population has several distinct local populations with more or less reduced gene flow between them. If one of these local populations is subjected to increased input of mutations, mutants may migrate over the total population via open channels of gene flow. However, if elevated mutation input over several generations is accompanied by increasing sterility, this results in a kind of "semipermeable" isolation to the affected local population i.e., no emigration of genetic information due to sterility, but immigration of unmutated genotypes. Dependent on the duration of environmental mutagen input and rate of immigration, either this isolation can speed up selective adaptation (see *RÖ*-populations) or immigrants can repopulate a deserted local niche.

In the above sections, there are two cases of superficial concordance with the sterility in ENU populations: (1) active transposons like P-elements induce mutations *and* sterility by hybrid dysgenesis, and (2) genetic factors like *av* can be triggered by mutagen input above a threshold level to sterilize the mutagen-treated individual via malfunction of the gonads. In spite of all discrepancies at closer look (hybrid dysgenesis only in first-generation hybrids; successful selection against *av*, but accumulation of ENU sterility), there are several genetically controlled mechanisms in a population that can trigger sterility on mutation input. These may well be part of unspecific "shock reactions" that, according to McClintock (1984), are a general feature of a genome. In any case, they can be used to maintain a mutagen-mutation equilibrium (Fig. 7).

6 Conclusion

Mutagenesis in living systems is as old as life and evolution. In fact, mutagenesis is essential for evolution. In adapted organisms new mutants are seldom positive; they may sometimes be neutral, but are quite often detrimental. Mutation rates should therefore be as high as necessary to enable further adaptive evolution, and as low as possible to avoid excess in genetic load. In other words, selection should maintain optimal mutation rates for given populations and their environments. This means regulation of intrinsic mutation rates according to genetic variability available within the gene pool of a local population, and to adaptive constraints acting on this gene pool. These constraints include environmental mutagens. The irradiated *RÖ*-populations of *Drosophila melanogaster* are an example of such an adaptation of populations to increased input of mutagens. It is achieved by various means. One main factor is the fine-tuned regulation of mutation rates as expected from the hypothesis of mutagen-mutation equilibria in evolution. As became more evident from studies of the ENU-populations, mutagen-induced sterility in local populations (not only by chromosomal aberrations, but by accumulation of effects possibly due to polygenic mutations) contributes to such equilibria within species. However, in the long run, the regulation of mutation rates according to environmental requirements is the crucial point. Evidence from the *RÖ*-populations indicates that this regulation may well be achieved by movable genetic elements. A central role of transposons in regulation of mutation rates, therefore, needs increased attention, because they may be more important than environmental mutagens themselves. Likewise, adaptive regulation and, hence, genetic control of mutation rates, at least in the germ line, points to the importance of individual genotypes in mutagenesis. Their evaluation may be of higher relevance in genetic risk estimates than that of weak environmental mutagens.

Acknowledgements. Technical assistance by Petra Goldbeck, Ingeborg Grade, and Susanne Hoffmann in part of recent work reported here, is gratefully acknowledged. It is a pleasure to thank Nebile Çürüksu for her generous permission to make use of some unpublished data on the ENU populations. These studies were supported by grant No. 150/2-1 from the Deutsche Forschungsgemeinschaft.

References

Baltimore D (1981) Somatic mutation gains its place among the generators of diversity. Cell 26: 295–296

Belyaeva ES, Pasyukova EG, Gvozdev VA, Ilyin YV, Kaidanov LZ (1982) Transposition of mobile dispersed genes in *Drosophila melanogaster* and fitness of stocks. Mol Gen Genet 185: 324–328

Benado MB, Ayala FJ, Green MM (1976) Evolution of experimental "mutator" populations of *Drosophila melanogaster*. Genetics 82: 43–52

Biémont C (1986) Polymorphism of the mdg-1 and I mobile elements in *Drosophila melanogaster*. Chromosoma (Berl) 93: 393–397

Biémont C, Aouar A (1987) Copy-number dependent transpositions and excisions of the mdg-1 mobile element in inbred lines of *Drosophila melanogaster*. Heredity 58: 39–47

Biémont C, Aouar A, Arnault C (1987) Genome reshuffling of the copia element in an inbred line of *Drosophila melanogaster*. Nature (Lond) 329: 742–744

Bingham PM, Kidwell MG, Rubin GM (1982) The molecular basis of P-M hybrid dysgenesis: The role of the P element, a P-strain-specific transposon family. Cell 29: 995–1004

Campbell A (1983) Transposons and their evolutionary significance. In: Nei M, Koehn RK (eds) Evolution of genes and proteins. Sinauer, Sunderland, pp 258–279

Chao L, Vargas C, Spear BB, Cox EC (1983) Transposable elements as mutator genes in evolution. Nature (Lond) 303: 633–635

Charlesworth B, Charlesworth D (1983) The population dynamics of transposable elements. Genet Res 42: 1–27

Cox EC, Gibson TC (1974) Selection for high mutation rates in chemostats. Genetics 77: 169–184

Dobzhansky T, Spassky B, Spassky N (1952) A comparative study of mutation rates in two ecologically diverse species of *Drosophila*. Genetics 37: 650–664

Doerjer F, Bedell MA, Oesch F (1984) DNA adducts and their biological relevance. In: Obe G (ed) Mutations in man. Springer, Berlin Heidelberg New York Tokyo

Engels WR (1984) A trans-acting product needed for P factor transposition in *Drosophila*. Science 226: 1194–1196

Engels WR, Preston CR (1984) Formation of chromosome rearrangements by P factors in *Drosophila*. Genetics 107: 657–678
Fitzpatrick BJ, Sved JA (1986) High levels of fitness modifiers induced by hybrid dysgenesis in *Drosophila melanogaster*. Genet Res 48: 89–94
Foster PL, Davis EF (1987) Loss of an apurinic/apyrimidinic site endonuclease increases the mutagenicity of N-methyl-N'-nitro-N-nitrosoguanidine to *Escherichia coli*. Proc Natl Acad Sci USA 84: 2891–2895
Gerasimova TI, Mizrokhi LJ, Georgiev GP (1984) Transposition bursts in genetically unstable *Drosophila melanogaster*. Nature (Lond) 309: 714–716
Gerasimova TI, Matjunina LV, Mizrokhi LJ, Georgiev GP (1985) Successive transposition explosions in *Drosophila melanogaster* and reverse transpositions of mobile dispersed genetic elements. EMBO J 4: 3773–3779
Gillespie JH (1981) Mutation modification in a random environment. Evolution 35: 468–476
Ginzburg LR, Bingham PM, Yoo S (1984) On the theory of speciation induced by transposable elements. Genetics 107: 331–341
Golub ES (1987) Somatic mutation: diversity and regulation of the immune repertoire. Cell 48: 723–724
Holsinger KE, Feldman MW (1983) Modifiers of mutation rate: Evolutionary optimum with complete selfing. Proc Natl Acad Sci USA 80: 6732–6734
Ives PT (1950) The importance of mutation rate genes in evolution. Evolution 4: 236–252
Kidwell MG, Kidwell JF, Ives PT (1977a) Spontaneous non-reciprocal mutation and sterility in strain crosses of *Drosophile melanogaster*. Mutat Res 42: 89–98
Kidwell MG, Kidwell JF, Sved JA (1977b) Hybrid dysgenesis in *Drosophila melanogaster:* a syndrome of aberrant traits including mutation, sterility, and male recombination. Genetics 86: 813–833
Kimball RF (1987) The development of ideas about the effect of DNA repair on the induction of gene mutations and chromosomal aberrations by radiation and by chemicals. Mutat Res 186: 1–34
King RC (1970) Ovarian development in *Drosophila melanogaster*. Academic Press, London
Lindsley DL, Grell EH (1968) Genetic variations of *Drosophila melanogaster*. Carnegie Inst Wash Publ No 627
Loeb LA, Preston BD (1986) Mutagenesis by apurinic/apyrimidinic sites. Annu Rev Genet 20: 201–230
Mackay TFC (1986) Transposable element-induced fitness mutations in *Drosophila melanogaster*. Genet Res 48: 77–87
McClintock B (1984) The significance of responses of the genome to challenge. Science 226: 792–801
Meyer J, Jäck HM, Ellis N, Wabl M (1986) High rate of somatic point mutation in vitro in and near the variable-region segment of an immunoglobulin heavy chain gene. Proc Natl Acad Sci USA 83: 6950–6953
Mukai T (1964) The genetic structure of natural populations of *Drosophila melanogaster*. I. Spontaneous mutation rate of polygenes controlling viability. Genetics 50: 1–19
Mukai T (1979) Polygenic mutation. In: Thompson JN, Thoday JM (eds) Quantitative genetic variation, Academic Press, Lond, pp 177–196
Mukai T, Chigusa SI, Mettler LE, Crow JF (1972) Mutation rate and dominance of genes affecting viability in *Drosophila melanogaster*. Genetics 72: 335–355
Nöthel H (1969) Genetische Konstitution und Strahlenempfindlichkeit. Habilitationsschrift, Freie Universität Berlin
Nöthel H (1983) Investigations on radiosensitive and radioresistant populations of *Drosophila melanogaster* XVI. Adaptation to the mutagenic effects of X-rays in several experimental populations with irradiation histories. Mutat Res 111: 325–340
Nöthel H (1987) Adaptation of *Drosphila melanogaster* populations to high mutation pressure: evolutionary adjustment of mutation rates. Proc Natl Acad Sci USA 84: 1045–1049
Nöthel H, Weber M (1976) Investigations on radiosensitive and radioresistant populations of *Drosophily melanogaster* VII. High relative radioresistance to the induction of sex-linked recessive lethals in stage-7 oocytes of RÖ I_4. Mutat Res 36: 245–248

O'Hare K, Rubin GM (1983) Structures of P transposable elements and their sites of insertion and excision in the *Drosophila melanosgaster* genome. Cell 34: 25–35
Radman M, Wagner R (1986) Mismatch repair in *Escherichia coli*. Annu Rev Genet 20: 523–538
Ramel C (1983) Polygenic effects and genetic changes affecting quantitative traits. Mutat Res 114: 107–116
Rio DC, Laski FA, Rubin GM (1986) Identification and immunochemical analysis of biologically active *Drosophila* P element transposase. Cell 44: 21–32
Rubin GM, Kidwell MG, Bingham PM (1982) The molecular basis of P-M hybrid dysgenesis: the nature of induced mutations. Cell 29: 987–994
Rusell WL, Hunsicker PR, Raymer GD, Steele MH, Stelzner KF, Thompson HM (1982) Dose-response curve for ethylnitrosourea-induced specific-locus mutations in mouse spermatogonia. Proc Natl Acad Sci USA 79: 3589–3591
Sankaranarayanan K, Sobels FH (1976) Radiation genetics. In: Ashburner M, Novitski E (eds) The genetics and biology of *Drosophila*. Academic press, Lond, vol 1c pp 1089–1250
Sturtevant AH (1937) Essays on evolution. I. On the effects of selection on mutation rate. Q Rev Biol 12: 464–467
Syvanen M (1984) The evolutionary implications of mobile genetic elements. Annu Rev Genet 18: 271–293
Voelker RA, Greenleaf AL, Gyurkovics H, Wisely GB, Huang SM, Searles LL (1984) Frequent imprecise excision among reversions of a P element-caused lethal mutation in *Drosophila*. Genetics 107: 279–294
Vogel E, Natarajan AT (1979) The relation between reaction kinetics and mutagenic action of mono-functional alkylating agents in higher eukayotic systems. I. Recessive lethal mutations and translocations in *Drosophila*. Mutat Res 62: 51–100
Vogel EW, Dusenbery RL, Smith PD (1985) The relation between reaction kinetics and mutagenic action. IV. The effects of the excicion-defective mei-9_{L1} and mus (2)201_{D1} mutants on alkylation-induced genetic damage in *Drosophila*. Mutat Res 149: 193–207
Williams LD, Shaw BR (1987) Protonated base pairs explain the ambiguous pairing properties of O^6-methylguanine. Proc Natl Acad Sci USA 84: 1779–1783
Yarosh DB (1985) The role of O^6-methylguanine-DNA methyltransferase in cell survival, mutagenesis and carcinogenesis. Mutat Res 145: 1–6

Development and Perspectives of the Human Lymphocyte Micronucleus Assay

R. HUBER and M. BAUCHINGER[1]

Contents

1 Introduction

The occurrence of micronuclei (MN) in plant cells after exposure to chemical substances and ionizing radiation has been a well-known phenomenon for several decades. These MN are smaller, accessory nuclear bodies in the cytoplasm which appear in addition to the main nucleus. They are believed to result primarily from acentric chromatin elements that were not incorporated into daughter nuclei during cell division. They may, however, also arise from whole chromosomes that failed to participate in normal anaphase movement.

Between 1973 and 1975 the so-called micronucleus test was developed for mammalian polychromatic erythrocytes from bone marrow (Boller and Schmid 1970; Ledebur and Schmid 1973; Schmid 1973; 1975). The mouse bone marrow MN test is now widely used as the primary short-term in vivo genotoxicity assay (Ashby et al. 1983).

Strategies of mutagenicity testing always include an in vitro chromosome assay. However, the classical metaphase chromosome analysis is very time-consuming and requires high practical expertise. In contrast, scoring of MN is a simple and rapid method. Therefore, the adaptation of the MN assay to the culture system of human lymphocytes offers obvious advantages for the evaluation of chromosomal aberrations. Once sufficiently established, the MN assay system might be usefully applied as a supplementary screening technique for detecting chromosome damage induced in vivo by chemicals and ionizing radiation in even larger populations. We report here on the advances in improving the technique as well as on its practical applications.

[1] Institut für Strahlenbiologie, Gesellschaft für Strahlen- und Umweltforschung, Ingolstädter Landstraße 1, D-8042 München-Neuherberg

2 Lymphocyte Preparation with Destroyed Cytoplasm

Countryman and Heddle (1976) and Countryman et al. (1977) described a method for the demonstration of radiation-induced MN in human lymphocyte preparations. Due to the hypotonic treatment of the cells, this technique yields mainly lymphoblast nuclei without recognizable cytoplasm (Fig. 1a). The MN are more or less separated from these nuclei. To provide a mutual association of MN with main nuclei, auxiliary criteria such as the location of MN within three or four nuclear diameters of a nucleus should be adopted.

It was observed that the frequency of MN increased with increasing culture time and that maximal MN counts were obtained at 84 or 96 h (Countryman and Heddle 1976). The first dose-response curve published for MN induced by X-rays, following exposures of 50–400 R, was fitted by the traditional power law model $Y = kD^n$, with a dose exponent $n = 1.2$, (Y = MN frequency, D = radiation dose in R, and k = a constant). This is in contrast to the exponent values of 1.8–2.0 from dose response curves for two-break aberrations induced by sparsely ionizing radiations which are scored in metaphase analyses (Evans and O'Riordan 1975). The differences were attributed to the influence of radiation-induced delays in cell division and to increased cell death at higher doses on the expression of MN.

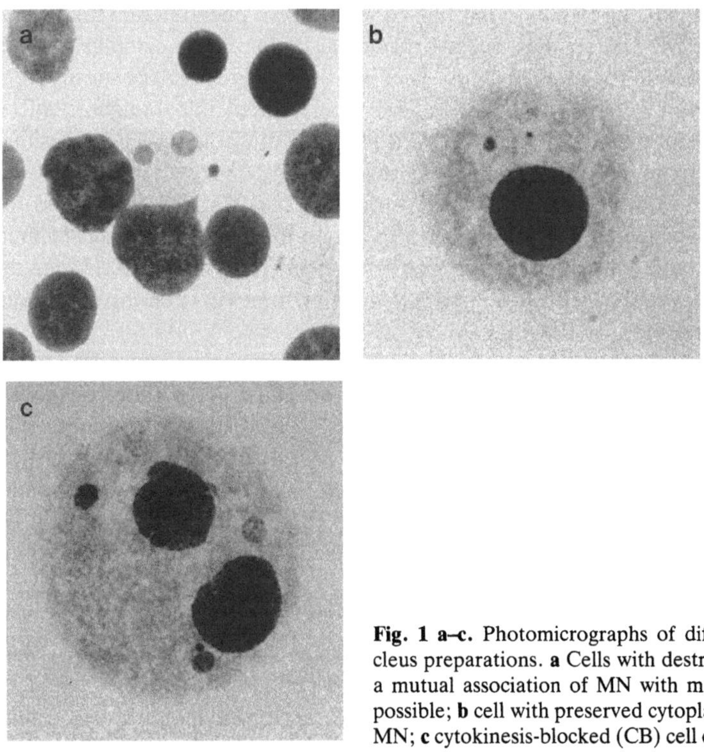

Fig. 1 a–c. Photomicrographs of different micronucleus preparations. **a** Cells with destroyed cytoplasm; a mutual association of MN with main nuclei is impossible; **b** cell with preserved cytoplasm containing 2 MN; **c** cytokinesis-blocked (CB) cell containing 5 MN

A similar influence on cell proliferation was evident in a later experiment with ^{137}Cs-gamma radiation at doses of 10–400 rad (Krepinski and Heddle 1983). There was little increase in the MN frequency after 400 rad compared to 300 rad. For 10–300 rad the dose-response curve appeared to be linear-quadratic. Regarding the application of the MN assay for estimating individual radiation exposure, the authors suggested that doses below 25 rad cannot be reliably detected. A comparison of chromosome analysis and MN counts from blood samples of four nuclear power plant workers chronically exposed to mixed gamma and tritium radiation revealed only a poor resolving power of the MN assay for these low level exposure conditions.

Split-dose experiments with two doses of 200 R (Countryman and Heddle 1976) revealed the typical fractionation effect as has been observed by several groups for two-hit aberrations in studies of radiation-induced chromosome damage. The MN frequency was reduced by a factor of about 2 and the rejoining time was measured to be between 30 and 60 min.

It could also be demonstrated that the MN assay is a suitable means to detect increased sensitivity of lymphocytes to 50–400 R of X-rays in Down's syndrome (Countryman and Heddle 1976; Countryman et al. 1977) and to Mitomycin C in Fanconi's anemia, as well as of ataxia telangiectasia fibroblast cells to gamma irradiation (Heddle et al. 1978).

Norman et al. (1978) applied the MN assay to test the effect of chemicals on human lymphocytes. They observed increased MN frequencies in blood samples from nine patients treated with diatrizoate contrast medium for angiography and in vitro due to triiodobenzoic acid derivatives. The latter observation confirms earlier results derived with the MN assay from root meristems of *Allium cepa* (Bauchinger 1964) and Chinese hamster cells (Schmid and Bauchinger 1976) treated with X-ray contrast media.

In a subsequent study of 73 young adults from 18–47 years, practical consequences became apparent from these results (Norman et al. 1984). An increased frequency of MN was observed with donor age, whereas sex and low-dose medical X-ray examination 2 years before blood sampling had no significant effect. However, a small, statistically significant increase of MN counts was revealed in patients who had an X-ray examination that involved the injection of contrast media into the blood. Unfortunately, no information on the donors' smoking habits were given.

Norman et al. (1985) carried out a study to evaluate whether the MN assay can be used to identify chromosome damage in human populations. Chromosome aberrations and MN were analyzed in a learning set and a test set, consisting of samples from a number of people occupationally exposed to ionizing radiatons or to other clastogens, and of therapeutically irradiated patients. It was demonstrated that people with high frequencies of aberrations, defined as one or more cells with unstable aberrations per 100–200 metaphases, could be correctly identified by means of MN counts. As in the previous study, age was a significant factor in MN counts. Taking this into consideration, the MN assay should provide a method for screening human populations for chromosome aberrations.

The MN assay was also used in experimental mutagenesis model systems, e.g. to assess interactions of mutagenic agents with respect to repair and cell sensitivity. Boyes and Koval (1983) observed significantly reduced incidences of MN in lymphocytes cultured for 72–120 h and irradiated in G_0 with 50–800 R gamma radiation, then post-treated with 1 mM caffeine. In contrast, irradiation of PHA-stimulated cultures at 48 h and caffeine post-treatment greatly enhanced the MN frequency after radiation doses greater than 200 R at a culture time of 72 h.

3 Lymphocyte Preparation with Preserved Cytoplasm

With the conventional method of lymphocyte preparation the cell cytoplasm is destroyed during hypotonic treatment. Since, according to Iskandar (1979), with this technique the nuclei tend to clump, he presented a method by which the cytoplasm could be preserved (Fig.1b). This could be achieved by changing the osmotic pressure of the hypotonic solution from about 77 mOsmol as used in conventional techniques up to 285 m Osmol which is nearly the concentration of an isotonic solution. Huber et al. (1983) used 200 mOsmol in combination with careful freezing of fixed cells onto the slides. Högstedt (1984) suggested a method without hypotonic treatment of cells smeared on slides.

With a preserved cytoplasm a more precise identification of MN can be derived. Now that MN can be easily associated with the main nucleus, their distribution among cells can be also analyzed (Huber et al. 1983). The modified MN technique was used in a series of studies for the detection of the effect of mutagenic clastogens in vitro and in vivo.

No increase in MN frequency could be observed in workers exposed to ethylene oxide (Högstedt et al. 1983a), but a small increase was detected in the case of styrene exposure (Högstedt et al. 1983b). Inconsistent results were reported on the effect of cigarette smoking on MN counts. Significantly higher frequencies were found in blood samples from smokers compared with non-smokers (Högstedt et al. 1983a; Stenstrand 1985), while Obe et al. (1982) observed no significant differences between these two groups. Since smoking had significant effects on the yields of structural chromosome aberrations, the authors believe that MN analysis seems not to be sensitive enough to detect differences that can be seen with chromosome analysis.

The clastogenic effect induced in vivo in human lymphocytes by metabolically activated Aflatoxin B_1 (Iskandar and Vijayalaxmi 1981),vinyl acetate (Mäki-Paakkanen and Norppa 1987), bleomycin, and 0.1–1.0 Gy ^{137}Cs gamma irradiation (Aghamohammadi et al. 1984; Cole et al. 1982) could be detected with the MN assay. Both these latter studies were carried out with cord blood lymphocytes, for which a very low background frequency of 3 MN per 1000 cells was found.

Fenech and Morley (1985c) analyzed the effect of donor age on spontaneous MN frequencies. A positive correlation was found with an approximately fourfold increase in MN counts in cultures at both 72 and 96 h from 80-year-old donors compared to cultures from newborns. An unexpected observation was that

X-irradiation in the exposure range from 50–400 R in vitro produced a significantly lesser increase in MN fequency in lymphocytes from older individuals than from young individuals. This result was attributed to reduced capacities of T-lymphocytes from elderly donors responding to mitogenic stimulation and proliferating.

Huber et al. (1983) investigated the suitability of the MN assay for biological dosimetry of radiation exposure. After irradiation of whole blood with 220 kV X-rays, the dose-response relationships for MN were analyzed from preparations with destroyed and with preserved cytoplasm. The shapes of the dose-effect curves at different culture periods derived with the first method were not uniform and revealed large variations (Fig. 2). With the second method, approximately five to six times more MN were demonstrable and clear dose-response relations could be described (Fig. 3). Maximal MN counts were obtained at 60 h. Due to the preserved cytoplasm, the intercellular distribution of MN could be evaluated (Table 1). As in the case of structural chromosomal changes, such information is necessary for the derivation of appropriate statistical weights (reciprocal sample variance), which should be used to obtain reliable regression analyses. At the dose range of 0–4 Gy, the data could be fitted by either the linear model $Y = b(0) + b(1)D$ or the linear-quadratic model $Y = b(0) + b(1)D + b(2)D^2$, where Y = MN frequency per cell, b(0), b(1) and b(2) are estimated parameters for spontaneous and induced MN respectively, and D = dose. At 48 h, data could be described by the linear model, while at 84 and 96 h, by the linear-quadratic model. At 60 and 72 h, no such definite conclusions could be drawn.

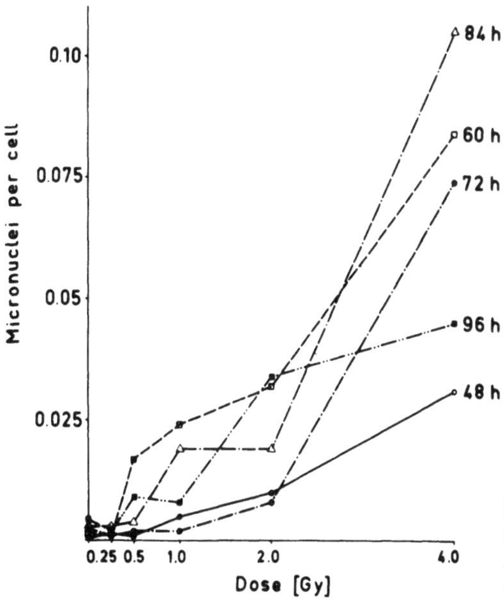

Fig. 2. Dose-effect curves for X-ray induced MN at different culture times. Data derived from preparations with destroyed cytoplasm (Huber et al. 1983)

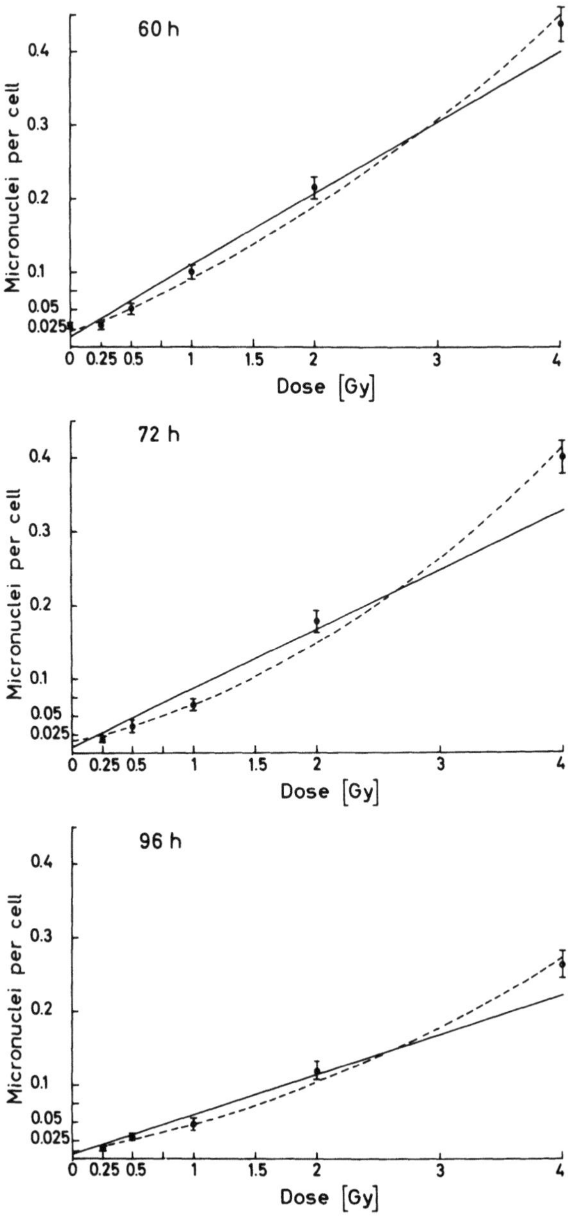

Fig. 3. Linear (*full line*) and linear-quadratic (*broken line*) dose-effect curves for X-ray induced MN at different culture times. Data derived from preparations with preserved cytoplasm (Huber et al. 1983)

Table 1. Frequency distribution of micronuclei among cells after X-irradiation. Data were derived at different culture times from preparations with preserved cytoplasm (Huber et al. 1983)

Culture time (h)	Dose (Gy)	Number of cells with 'n' micronuclei					Micronuclei	
		0	1	2	3	4–5	Total number	Mean number per cell
60	0	1951	44	3	2	–	56	0.028
	0.25	1953	40	7	–	–	54	0.027
	0.5	957	38	4	1	–	49	0.049
	1.0	910	82	6	2	–	100	0.100
	2.0	815	156	28	1	–	215	0.215
	4.0	683	220	75	19	3	439	0.439
72	0	1959	38	3	–	–	44	0.022
	0.25	1969	26	5	–	–	36	0.018
	0.5	966	32	2	–	–	36	0.036
	1.0	942	52	6	–	–	64	0.064
	2.0	847	131	18	4	–	179	0.179
	4.0	707	214	57	17	5	400	0.400
96	0	1972	28	–	–	–	28	0.014
	0.25	1974	25	1	–	–	27	0.014
	0.5	972	26	2	–	–	30	0.030
	1.0	956	41	3	–	–	47	0.047
	2.0	899	85	14	2	–	119	0.119
	4.0	798	153	38	8	3	265	0.265

At each culture period the background frequency of MN amounted to values observed at the lowest dose of 0.25 Gy. Thus, despite the possibility to establish a clear dose-effect relation for the induction of MN in vitro, the assay system does not seem sensitive enough to detect a significant increase in the incidence of MN after exposure to doses below 0.3 Gy.

4 Modified Harlequin Staining Procedure

MN which are equivalents of acentric fragments, or of lagging chromosomes not incorporated into a daughter nucleus during mitosis, cannot be detected until the first post-treatment interphase. This means that the expression of MN is highly dependent on cell proliferation kinetics. Thus, the proportion of dividing lymphocytes in culture must depend on the extent of the response to the mitogen. Interindividual or intraindividual variations for this process are well known. Any inhibitory influence on cell proliferation will also affect the proportion of dividing cells. Therefore, it is not surprising that a standardization of culture time to obtain actual maximal response after treatment with clastogens is very difficult.

To overcome these difficulties Pincu et al. (1984) developed a modification of the FPG (fluorescence plus Giemsa) harlequin staining technique (Perry and Wolff 1974) that provides easy identification of proliferating lymphocytes, which have blue nuclei, from non-proliferating cells with red nuclei. When MN are scored only in cells having blue nuclei, those cells that have not divided at least once in culture should, therefore, be excluded from the assay. The differential staining is achieved by a preceding continuous incorporation of high concentrations of 5-bromodeoxyuridine (BrdU) (4×10^{-4} M).

A comparison of MN counts in blue cells and those from all cells in cultures given 0.25–8.0 Gy of ^{60}Co gamma rays revealed a dose-dependent increase of the MN yields. The two-fold increase at lower doses and the four-fold increase at doses greater than 3.0 Gy thus indicated the existence of a large fraction of non-proliferating cells.

The results demonstrate that with the application of the harlequin staining procedure, the cell kinetic problems in the MN assay can be reduced. However, for a standardization of this technique it should be investigated whether the MN counts could be influenced by the following factors.

One source of error might be the high BrdU concentration of 4×10^{-4} M. Though Pincu et al. (1984) found no significant effects of this concentration on MN yields, others observed the induction of MN by chromosome breaks resulting from BrdU (Fenech and Morley 1985b; Raj and Heddle 1980). Boyes and Koval (1985) showed that BrdU concentrations up to about 10^{-4} M had no significant effect on MN frequency. Post-treatment of ^{137}Cs gamma-irradiated lymphocytes in doses ranging from 0.5–8.0 Gy with 10^{-4} M BrdU decreased the MN frequency at doses greater than 0.5 Gy. The mitotic index was also significantly decreased at 2.0–8.0 Gy. The inhibitory effect of BrdU on lymphocyte proliferation in culture is also well documented from the analyses of harlequin-stained metaphase preparations (Kolin-Gerresheim and Bauchinger 1981; Scott and Lyons 1979). Thus, particularly the possibility of interaction of BrdU with a clastogen could influence the quantitative analysis of MN induction, and should be clarified in more detail.

Fenech and Morley (1985b) noted a further drawback of this modified MN assay, in that cells in late S or G_2 of the first cycle are scored in addition to cells in the second cycle. To solve the kinetic problem in the MN assay, they suggested four new methods to identify MN in lymphocytes that have divided only once. Of these, the autoradiographic method and the cytokinesis-block method will be discussed here.

5 Autoradiographic Method

With this approach separated lymphocytes are pulse-labeled with ^{3}H-thymidine 48 h after PHA stimulation. Cells which incorporated tritiated thymidine in the S phase can be identified after cell division by their labeled nuclei. Scoring of MN is thus confined to labeled cells that have divided only once at 72–84 h (Fenech and Morley 1985a, b). The induction of MN was clearly dose-dependent when cells were exposed to 100 and 200 rad of 100 kV X-rays. The autoradiographic method cannot be used for a reliable analysis of baseline frequencies, because chromosome aberrations, and thus also MN, are induced by β-rays emitted from tritium. As a routine technique the method is too laborious.

6 Cytokinesis-Block Method

With this approach Cytochalasin B (Cyt-B) is added to separated lymphocytes or whole blood 44–48 h after PHA stimulation.

Cytokinesis is blocked by Cyt-B, whereas the nuclear division is not inhibited (Carter 1967; 1972; Kelly and Sambrock 1973; O'Neill 1972; Ridler and Smith 1968; Wright and Hayflick 1972). Cytokinesis-blocked (CB) cells accumulate in their first division cycle and can be identified by their binucleate appearance at 72 h (Fenech and Morley, 1985a; b) (Fig.1c). These CB cells have undergone a nuclear division and MN can be easily recognized.

A comparison of the data obtained by the autoradiographic method and the CB method for the number of MN induced by 100 and 200 rad of 100 kV X-rays revealed close agreement (Fenech and Morley 1985b). Further dose-response experiments carried out with the CB method (Fenech and Morley 1985b, 1986) yielded approximately linear relationships for high-dose (100–400 rad) and for low-dose 100 kV X-irradiation (5–40 rad). The MN frequencies in cultures given 5 rad were significantly higher than in control cultures, with an average of only 4.4 ± 2.6 MN per 500 CB cells.

Mitchell and Norman (1987) obtained non-linear dose-response curves for the yield of MN induced by 0.05–0.5 Gy of ^{60}Co gamma rays and 90 kVp X-rays with significantly greater slopes of the curves at doses above 0.15 Gy. Correlation coefficients significantly greater than zero could be derived when the dose range between 0–0.15 Gy was analyzed separately. A curve for ^{60}Co gamma rays at doses of 0.05–0.8 Gy, derived by using the modified harlequin method, had a similar shape. However, no significant correlation could be demonstrated for the low-dose region. The discrepancy was attributed to differences in cell proliferation kinetics in the two assays. At 72 h the average number of nuclear divisions was 2.6 for blue cells (without Cyt-B) and 1.3 for Cyt-B treated cells. Thus, in the former assay many cells should have divided more than once, which may explain the apparent lack of radiation sensitivity at low doses.

Ramalho et al. (1988) compared the yield of radiation-induced MN (150 kV X-rays) assessed by the CB method (50–300 rad) or by the modified harlequin staining procedure (100–400 rad). Chromosome aberrations were determined in parallel preparations. The CB method was found to be more efficient to detect induced fragments. In dose fractionation experiments it could be demonstrated that the yield of MN reflects acentric fragments associated with exchange type aberrations as well as excess acentrics. No marked differences were found in the MN frequencies of CB cells when 200 rad were fractionated with intervals of 60–120 min or when 400 rad were split into two fractions separated by 180 min.

The CB MN technique was examined in two studies with 250 kV X-rays for its potential application as a method of biological radiation dosimetry (Kormos

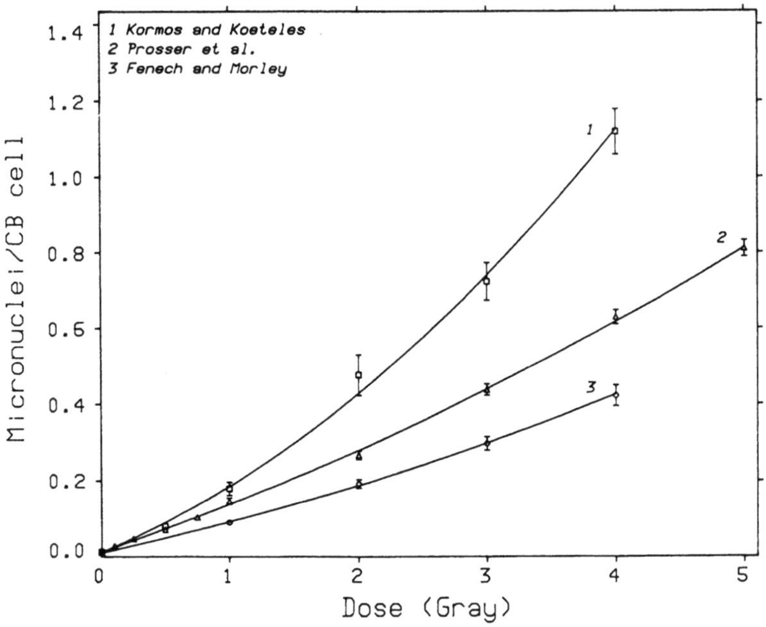

Fig. 4. Linear-quadratic dose-effect curves for MN induced by 250 kV (Data from Kormos and Köteles 1988 and from Prosser et al. 1988) and 100 kV X-rays (Data from Fenech and Morley 1985b) in CB cells

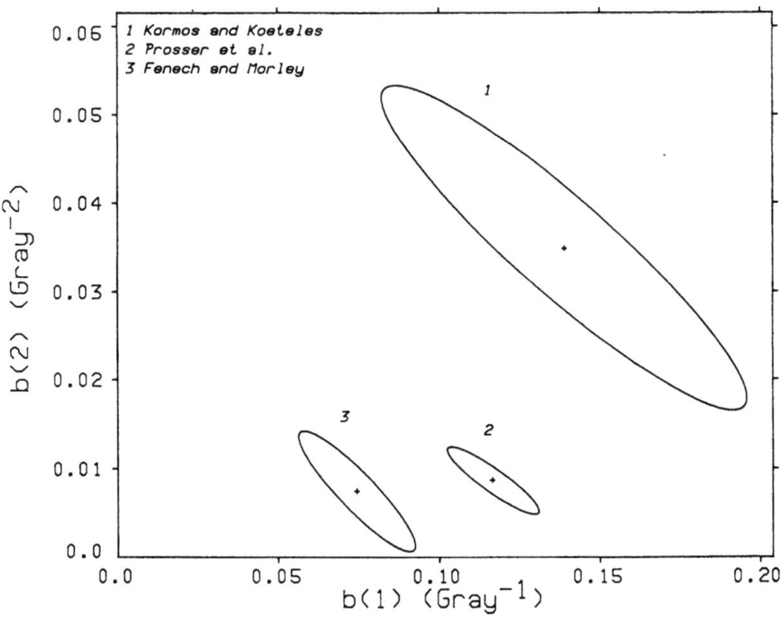

Fig. 5. 95% confidence regions for the parameters b(1) and b(2) of the linear-quadratic relations $Y = b(0) + b(1)D + b(2)D^2$ shown in Fig. 4

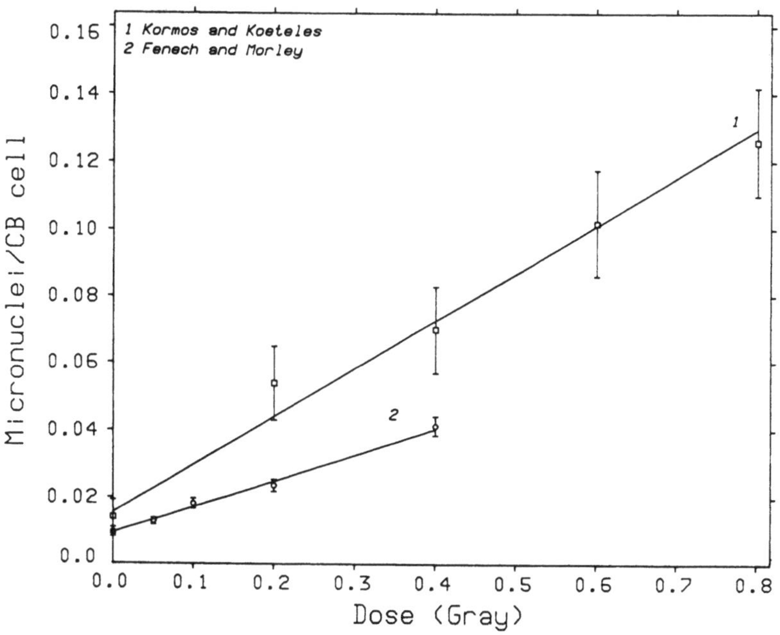

Fig. 6. Linear dose-effect curves for MN induced by 250 kV (Data from Kormos and Köteles 1988) and 100 kV X-rays (Data from Fenech and Morley 1986) in CB cells

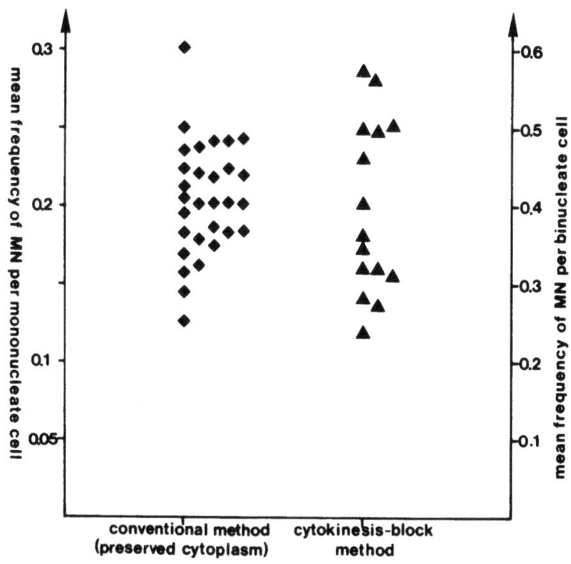

Fig. 7. Interindividual variation in MN yields of in vitro to 3 Gy gamma rays exposed blood samples from 30 donors assessed by a method with preserved cytoplasm (◆) and from 15 donors assessed by the CB method (▲) (Huber et al. 1989). For comparison of the two methods it has to be considered that a CB cell represents the equivalent of two daughter cells. Therefore, the ordinate scales for the frequencies of MN are differing by a factor of 2

and Köteles 1988; Prosser et al. 1988). For exposures in a high dose range the data have been fitted to the linear-quadratic model $Y = b(0) + b(1)D + b(2)D^2$ (Fig.4). For comparison we have included the 100 kV X-ray curve from Fenech and Morley (1985b) which can also be fitted to the quadratic model with a small but significant b(2) term. From the 95% confidence ellipsoids for the linear and the quadratic parameters of the corresponding dose-response relations, considerable differences are apparent between the different experiments (Fig.5). Similar differences were found for the linear dose-response curves obtained after exposures to 100 kV (Fenech and Morley 1986) and 250 kV (Kormos and Köteles 1988) X-rays in a low-dose range (Fig.6).

The observations of an age effect in vivo on baseline MN frequency, obtained in earlier studies with a conventional method (Fenech and Morley 1985c; Högstedt 1984); Norman et al. 1984, 1985), were confirmed with the CB method for a group of 42 individuals aged between 21–85 years (Fenech and Morley 1985c). Fenech and Morley conclude from their data (1985c, 1986) that the incremental effect is clearly underestimated by the conventional MN assay. With the CB method a rate of 4.3% increase per year was found. This numerical estimate is, however, not substantially different to 3.4 ± 1.3% increase per year in a group of 30 donors aged 20–52 years, studied with a conventional assay with preserved cytoplasm (Huber et al. 1989). In two further studies, no significant age effect was apparent when the CB method was applied (Huber et al. 1989; Prosser et al. 1988).

Using the CB method, significant interindividual variations could be detected for the baseline levels of MN and for the frequency of MN induced by in vitro irradiation of blood samples from 14 donors with 3.0 Gy ^{137}Cs gamma rays. Similar results were obtained with a conventional assay with preserved cytoplasm in which cell proliferation was observed in parallel cultures (Huber et al. 1989) (Fig.7).

7 Perspectives

Scoring of MN in human peripheral lymphocytes provides an easy and rapid method of measuring chromosomal damage. Since its first application to the lymphocyte culture system, the MN assay has been continuously improved. The most important advance is certainly the development of methods to account for the influence of cell proliferation kinetics in MN expression. Among the different approaches explored, the CB method has proven to be the most appropriate for quantitative analyses.

Still, an inherent problem of MN assays is that one cannot directly discriminate between MN containing only chromosomal fragments or whole chromosomes. Several attempts have been applied to overcome this problem either by measuring their sizes (Högstedt and Karlsson 1985; Wakata and Sasaki 1987; Yamamoto and Kikuchi 1980) or their DNA content (Heddle and Carrano 1977; Nüsse and Kramer 1984; Pincu et al. 1985). Evidence for the presence of whole chromosomes in MN of human lymphocytes came from the observation of chromocentres detected by C-banding (Banduhn and Obe 1985), and from the re-expression of metaphase chromosomes when MN of human and Chinese hamster cells were fused with whole cells (Viaggi et al. 1987).

The observation of Moroi et al. (1980) could be the decisive progress in the identification of centromeres in MN. Using anti-kinetochore antibodies from patients with the CREST variant of scleroderma, they showed that centromeres can be detected by immunofluorescence staining in interphase chromatin. Degrassi and Tanzarella (1988) used this approach in hamster cells and showed that about 80% of MN induced by either colchicine or chloral hydrate contained kinetochores, but only 9% of X-ray-induced MN were CREST positive.

Identical results are reported in CREST labelled CB human lymphoctes by Thomson and Perry (1988). Virtually all X-ray- and mitomycin-c-induced MN lacked kinetochore fluorescence, indicating that they originated from acentric chromosomal fragments. Of colcemid-induced MN, 79% revealed kinetochore fluorescence, reflecting their origin from whole chromosomes. Of baseline MN, 61% were CREST positive.

Thus a MN technique which combines a cytokinesis block and CREST immunofluorescence staining could be a valuable assay to interpret data on the clastogenic effect of a chemical substance, and to screen for aneuploidy-inducing agents. However, such an assay is not yet fully established. Reasons for this are discussed by Thomson and Perry (1988). Further experiments are required to elucidate whether the induction of chromosome aberrations by Cytochalasin B, the effect of specific kinetochore damage on CREST antibody binding or the loss of DNA from MN during the cell cycle may limit the application of this combined method.

Provided the precision and sensitivity of MN assays can be confirmed in further experiments and in vivo studies, they could be successfully applied for montoring larger populations exposed to ionizing radiation and chemical agents. In pre-employment examinations of occupational radiation exposures, MN scoring could be helpful for diagnosing potential radiosensitive individuals who are not obviously suffering from radiosensitive syndromes. Whether MN assays can be established as an additional biological routine method for estimating an individual's exposure to low doses of ionizing radiation cannot yet be decided from existing data.

A large interdonor variability observed for the baseline incidences of MN could impose some uncertainties on using this approach. An unequivocal detection of 0.05 Gy in vitro as suggested by Fenech and Morley (1986) does not necessarily hold for similar in vivo exposures. As recently discussed by Prosser et al. (1988) this may be the case only where a control unirradiated value for the particular donor is available. However, for accidental radiation exposure one would not have a pre-irradiation control sample. Thus, from existing data, the MN assay may be potentially useful to estimate acute exposures to low-LET radiation doses above 0.1–0.3 Gy.

Additional work is required to analyze the marked lack of correspondence between data on chromosomal aberrations, e.g. acentrics or total unstable aberrations and MN incidence (Prosser et al. 1988; Ramalho et al. 1988). A comparison with the respective dose-response relationships revealed a higher background frequency of MN, a higher linear coefficient and a smaller quadratic coefficient

(Prosser et al. 1988). Consequently, it should be analyzed whether the conventional dose-response models are also fully applicable to quantify MN data.

Finally, the value of MN assays for the rapid and precise detection of genotoxic effects could be further enhanced, if the initial attempts on scoring MN by computerised image analysis (Callisen and Norman, 1983; Fenech et al. 1988) or by flow cytometry (Nüsse and Kramer 1984) can be successfully extended.

Acknowledgement. We wish to thank H. Braselmann for discussion and statistical advice.

References

Aghamohammadi SZ, Henderson L, Cole RJ (1984) The human lymphocyte micronucleus assay. Response of cord blood lymphocytes to gamma irradiation and bleomycin. Mutat Res 130: 395–401

Ashby J, De Serres FJ, Draper MH, Ishidate M, Matter BE, Shelby M (1983) The two ICPS collaborative studies of short-term tests for genotoxicity and carcinogenicity. Mutat Res 109: 123–126

Banduhn N, Obe G (1985) Mutagenicity of methyl-2-benzimidazolecarbamate, diethylstilbestrol and estradiol: structural chromosomal aberrations, sister chromatid exchanges, C-mitoses, polyploidies and micronuclei. Mutat Res 156: 199–218

Bauchinger M (1964) Untersuchungen über die cytogenetische Wirkung von wasserlöslichen Röntgenkontrastmitteln. Z Zellforsch 63: 506–537

Boller K, Schmid W (1970) Chemische Mutagenese beim Säuger; Das Knochenmark des Chinesischen Hamsters als in-vivo-Testsystem. Hämatologische Befunde nach Behandlung mit Trenimon. Humangenetik 11: 35–54

Boyes BG, Koval JJ (1983) Clastogenic interactions of gamma radiation and caffeine in human peripheral blood cultures. Mutat Res 108: 239–249

Boyes BG, Koval JJ (1985) A cautionary note on the use of BrdU when determining micronucleus frequencies. Int J Radiat Biol 47: 341–342

Callisen H, Norman A (1983) Micronuclei of human lymphocytes as a biological dosimeter. In: Eisert WG, Mendelson ML (eds) Biological dosimetry: cytometric approaches to mammalian systems. Springer, Berlin Heidelberg New York, pp 171–179

Carter SB (1967) Effects of cytochalasins on mammalian cells. Nature (Lond) 213: 261–264

Carter SB (1972) The cytochalasins as research tools in cytology. Endeavour (Oxf) 31: 77–82

Cole RJ, Aghamohammadi SZ, Henderson L (1982) Frequencies of genetic lesions in lymphocytes from new-born humans. In: Sorsa M, Vainio H (eds) Mutagens in our environment. Alan Liss Inc, N Y pp 375–386

Countryman PI, Heddle JA (1976) The production of micronuclei from chromosome aberrations in irradiated cultures of human lymphocytes. Mutat Res 41: 321–332

Countryman PI, Heddle JA, Crawford E (1977) The repair of X-ray-induced chromosomal damage in trisomy 21 and normal diploid lymphocytes. Cancer Res 37: 52–58

Degrassi F, Tanzarella C (1988) Immunofluorescent staining of kinetochores in micronuclei: a new assay for the detection of aneuploidy. Mutat Res 203: 339–345

Evans HJ, O'Riordan (1975) Human lymphocytes for the analysis of chromosome aberrations in mutagen tests. Mutat Res 31: 135–138

Fenech M, Morley AA (1985a) Solutions to the kinetic problem in the micronucleus assay. Cytobios 43: 233–246

Fenech M, Morley AA (1985b) Measurement of micronuclei in lymphocytes. Mutat Res 147: 29–36

Fenech M, Morley AA (1985c) The effect of donor age on spontaneous and induced micronuclei. Mutat Res 148: 99–105

Fenech M, Morley AA (1986) Cytokinesis-block micronucleus method in human lymphocytes: effect of in vivo ageing and low dose X-irradiation. Mutat Res 161: 193–198

Fenech M, Jarvis LR, Morley AA (1988) Preliminary studies on scoring micronuclei by computerised image analysis. Mutat Res 203: 33–38

Heddle JA, Carrano AV (1977) The DNA content of micronuclei in mouse bone marrow by irradiation: evidence that micronuclei arise from acentric chromosomal fragments. Mutat Res 44: 63–69

Heddle JA, Benz RD, Countryman PI (1978) Measurement of chromosomal breakage in cultured cells by the micronucleus technique. In: Evans HJ, Lloyd DC (eds) Mutagen-induced chromosome damage in man. Edinburgh Univ Press, pp 191–200

Högstedt B (1984) Micronuclei in lymphocytes with preserved cytoplasm. A method for assessment of cytogenetic damage in man. Mutat Res 130: 63–72

Högstedt B, Karlsson A (1985) The size of micronuclei in human lymphocytes varies according to inducing agent used. Mutat Res 156: 229–232

Högstedt B, Gullberg B, Hedner K, Kolnig AM, Mitelman F, Skerfving S, Widegren B (1983a) Chromosome aberrations and micronuclei in bone marrow cells and peripheral blood lymphocytes in humans exposed to ethylene oxide. Hereditas 98: 105–113

Högstedt B, Akesson B, Axell K, Gullberg B, Mitelman F, Pero RW, Skerfving S, Welinder H (1983b) Increased frequency of lymphocyte micronuclei in workers producing reinforced polyester resin with low exposure to styrene. Scand J Work Environ Health 9: 241–246

Huber R, Streng S, Bauchinger M (1983) The suitability of the human lymphocyte micronucleus assay system for biological dosimetry. Mutat Res 111: 185–193

Huber R, Braselmann H, Bauchinger M (1989) Screening for interindividual differences in radiosensitivity by means of the micronucleus assay in human lymphocytes. Radiat Environ Biophys 28: 113–120

Iskandar O (1979) An improved method for the detection of micronuclei in human lymphocytes. Stain Technol 54: 221–223

Iskandar O, Vijayalaxmi (1981) The enhancement of the effect of Aflatoxin B_1 by metabolic activation with rat-liver microsomes on human lymphocytes assayed with the micronucleus test. Mutat Res 91: 63–66

Kelly F, Sambrock J (1973) Differential effect of Cytochalasin B on normal and transformed mouse cells. Nature (Lond) 242: 217–219

Kolin-Gerresheim I, Bauchinger M (1981) Dependence of the frequency of harlequin-stained cells on BrdU concentration in human lymphocyte cultures. Mutat Res 91: 251–254

Kormos C, Köteles GJ (1988) Micronuclei in X-irradiated human lymphocytes. Mutat Res 199: 31–35

Krepinski AB, Heddle JA (1983) Micronuclei as a rapid and inexpensive measure of radiation-induced chromosomal aberrations. In: Ishihara T, Sasaki MS (eds) Radiation-induced chromosome damage in man. Alan R Liss Inc, N Y pp 93–109

Ledebur M, Schmid W (1973) The micronucleus test; methodological aspects. Mutat Res 19: 109–117

Mäki-Paakkanen J, Norppa H (1987) Induction of micronuclei by vinyl acetate in mouse bone marrow cells and cultured human lymphocytes. Mutat Res 190: 41–45

Mitchell JC, Norman A (1987) The induction of micronuclei in human lymphocytes by low doses of radiation. Int J Radiat Biol 52: 527–535

Moroi Y, Peebles C, Fritzler MJ, Steigerwald J, Tan EM (1980) Autoantibody to centromere (kinetochore) in scleroderma sera. Proc Natl Acad Sci USA 77: 1627–1631

Norman A, Adams FH, Riley RF (1978) Cytogenetic effects of contrast media and triiodobenzoic acid derivatives in human lymphocytes. Radiology 129: 199–203

Norman A, Cochran S, Bass D, Roe D (1984) Effects of age, sex and diagnostic X-rays on chromosome damage. Int J Radiat Biol 46: 317–321

Norman A, Bass D, Roe D (1985) Screening human populations for chromosome aberrations. Mutat Res 143: 155–160

Nüsse M, Kramer J (1984) Flow cytometric analysis of micronuclei found in cells after irradiation. Cytometry 5: 20–25

Obe G, Vogt HJ, Madle S, Fahning A, Heller WD (1982) Double-blind study on the effect of cigarette smoking on the chromosomes of human peripheral blood lymphocytes in vivo. Mutat Res 92: 309–319
O'Neill FJ (1972) Chromosome pulverisation in cultured normal and neoplastic cells treated with cytochalasin B. J Natl Cancer Inst 49: 1733–1737
Perry P, Wolff S (1974) New Giemsa method for the differential staining of sister chromatids. Nature (Lond) 251: 156–158
Pincu M, Bass D, Norman A (1984) An improved micronucleus assay in lymphocytes. Mutat Res 139: 61–65
Pincu M, Callisen H, Norman A (1985) DNA content of micronuclei in human lymphocytes. Int J Radiat Biol 47: 423:432
Prosser JS, Moquet JE, Lloyd DC, Edwards AA (1988) Radiation induction of micronuclei in human lymphocytes. Mutat Res 199: 37–45
Raj AS, Heddle JA (1980) Simultaneous detection of chromosomal aberrations and sister-chromatid exchanges. Experience with DNA intercalating agents. Mutat Res 78: 253–260
Ramalho A, Sunjevaric I, Natarajan AT (1988) Use of the frequencies of micronuclei as quantitative indicators of X-ray-induced chromosomal aberrations in human peripheral blood lymphocytes: comparison of two methods. Mutat Res 207: 141–146
Ridler MAC, Smith GF (1968) The response of human cultured lymphocytes to cytochalasin B. J Cell Sci 3: 595–602
Schmid W (1973) Chemical mutagen testing on in vivo somatic mammalian cells. Agents Actions 3: 77–85
Schmid W (1975) The micronucleus test. Mutat Res 31: 9–15
Schmid E, Bauchinger M (1976) The cytogenetic effect of an X-ray contrast medium in Chinese hamster cell cultures. Mutat Res 34: 291–298
Scott D, Lyons CY (1979) Homogeneous sensitivity of human peripheral blood lymphocytes to radiation-induced chromosome damage. Nature (Lond) 278: 756–758
Stenstrand K (1985) Effects of ionizing radiation on chromosome aberrations, sister-chromatid exchanges and micronuclei in lymphocytes of smokers and nonsmokers. Hereditas 102: 71–76
Thomson EJ, Perry PE (1988) The identification of micronucleated chromosomes: a possible assay for aneuploidy. Mutagenesis 3: 415–418
Viaggi S, Bonatti S, Abbondandolo A (1987) New evidence for the presence of chromosomes in micronuclei of human Chinese hamster cells. Mutagenesis 2: 367–370
Wakata A, Sasaki MS (1987) Measurement of micronuclei by cytokinesis-block method in cultured Chinese hamster cells. Comparison with types and rates of chromosome aberrations. Mutat Res 190: 51–57
Yamamoto KI, Kikuchi Y (1980) A comparison of diameters of micronuclei induced by clastogens and by spindle poisons. Mutat Res 71: 127–131
Wright WE, Hayflick L (1972) Formation of anucleate and multinucleate cells in normal and SV40 transformed WI-38 by cytochalasin B. Exp Cell Res 74: 187–194

Detection of Nucleic Acids by Enzyme-Linked Immuno-Sorbent Assay (ELISA) Technique: An Example for the Development of a Novel Nonradioactive Labeling and Detection System with High Sensitivity*

C. KESSLER[1]

Contents

1 Previous Nonradioactive Nucleic Acid Detection Systems

Detection of minute quantities of the genetic material deoxyribonucleic acid (DNA) is a fundamental and fascinating issue in molecular biology.

The electron microscopy technique allowed direct analysis of complex structures – like those of human chromosomes (Sandberg 1980; Obe and Basler 1987). However, these optical methods do not allow the determination of the molecular structure of genetic material.

In addition to the development of potent methods for DNA sequencing (Maxam and Gilbert 1977, 1980; Sanger et al. 1977, 1980; Guo et al. 1983; Church and Gilbert 1984; Beck 1987; Wong et al. 1987; Hood et al. 1987; Kristensen et al. 1988; Toneguzzo et al. 1988), it was in particular the hybridization technique

[1] Boehringer Mannheim GmbH, Biochemical Research Center Penzberg, Department of Genetics, D-8122 Penzberg
* Dedicated to Dr. Felice A. de Jong

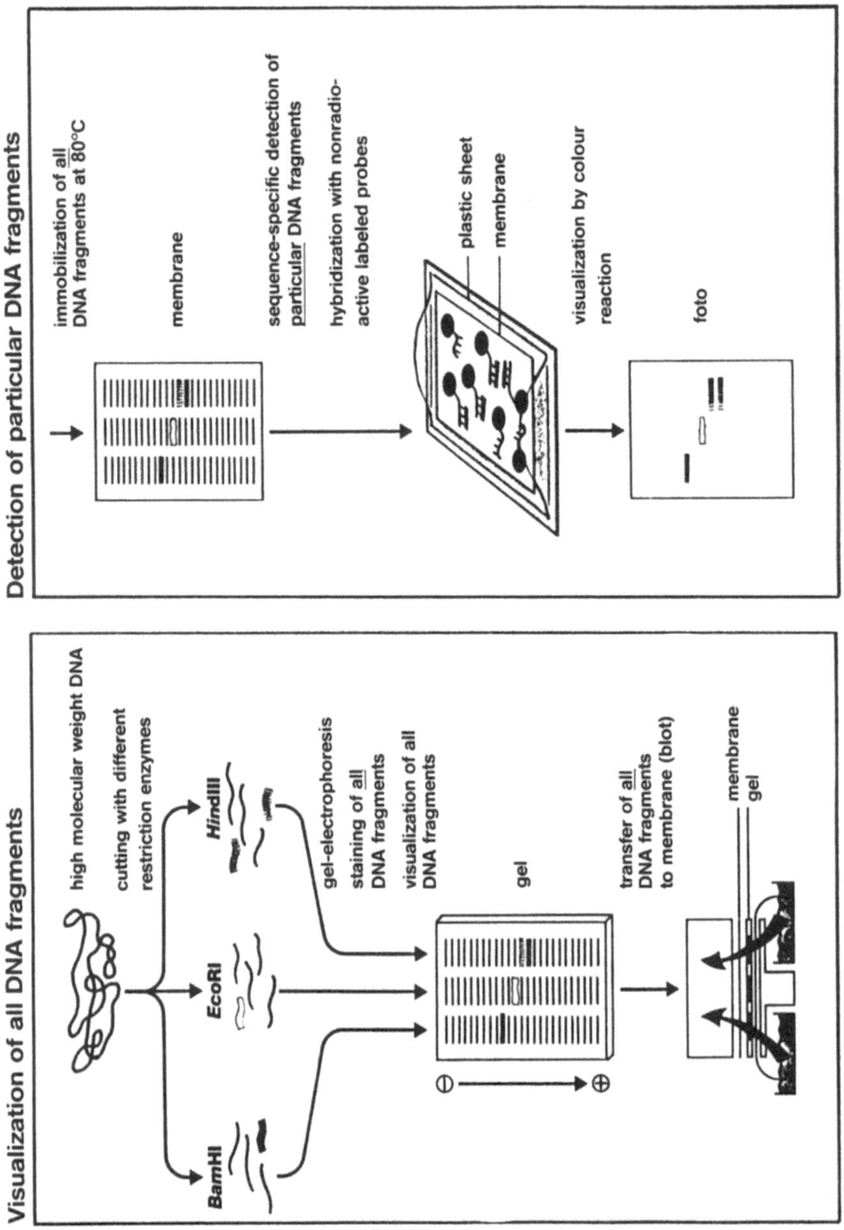

Fig. 1. Sequence-specific DNA detection by hybridization (Southern blotting)

(Gillespie and Spiegelman 1965; Denhardt 1966), as modified by E. Southern, which made it possible to detect and analyze certain DNA sections in a sequence-specific manner. This second fundamental technique for DNA analysis has been named "Southern blotting" in honor of its inventor and was adapted in later years as dot blots omitting DNA transfer (Southern 1975; Kafatos et al. 1979). Recently a method was described for direct hybridization in dried agarose gels without the need of DNA transfer (Mather 1988).

Sequence-specific DNA detection by Southern blotting and hybridization procedures with sequence-specific hybridization probes is performed as follows (Fig. 1): DNA is extracted from cells and cleaved by restriction enzymes (e.g. *Bam*HI, *Eco*RI or *Hin*dIII). The resulting fragments are separated according to size by electrophoresis on agarose gels. By appropriately arranging absorptive filter paper, the fragments are transfered from the gel onto the nitrocellulose filter. The result is a lift-off of the fragment pattern. After fixation of the fragment lift-off by baking the nitrocellulose filter at 80 °C, by UV crosslinking or by treatment with organic solvents sequence-specific detection of particular fragments is performed by hybridization with labeled DNA probes under stringent conditions (Meinkoth and Wahl 1984; Anderson and Young 1985). Specifically those fragments to which the DNA probe binds under stringent conditions are rendered visible by labeling the probe prior to hybridization. All other fragments remain invisible.

With an overall number of approx. 3×10^9 bp in the haploid human genome, detection of a 0.1–1 kB DNA section – hybridizing with labeled standard probes – requires a detection sensitivity of about $1 : 10^{-7}$. For the detection of single-copy genes with labeled standard probes, 0.1 pg of the homologous and thus hybridizing DNA section must be detectable when analyzing 1 μg genomic DNA. This is less than one trillionth of a gram.

Radioactively labeled probes have so far been used for DNA hybridization experiments (Maitland et al. 1987). The radioactive markers most frequently applied are [^{3}H], [^{32}P], [^{35}S] and [^{125}I] isotopes. For the detection of "single-copy" genes, [^{32}P]-isotopes are commonly used because their radiation intensity is very high. High rates of incorporation are achieved by homogeneous nick translation (Rigby et al. 1977; Telford et al. 1979) or random-primed labeling (Feinberg and Vogelstein 1983, 1984), both of which are catalyzed by the Klenow enzyme. 5'- or 3'-end-labeling with T4 polynucleotide kinase or terminal transferase results in lower rates of DNA labeling (Maxam and Gilbert 1980; Guo et al. 1983). The same holds true for filling in sticky ends with labeled deoxynucleotides and Klenow enzyme (Rigby et al. 1977).

However, the use of [^{32}P] isotopes has serious disadvantages:

1. The [^{32}P] isotope has a half-life of only 14.3 days (Baker 1969) and must therefore by used quickly;
2. In longer test series, newly labeled samples must be incorporated; this has a negative effect on the comparability of the test results;
3. In the case of older [^{32}P]-labeled samples, exposition periods of several days and sometimes even weeks are required;

4. Storage of [^{32}P]-labeled samples is therefore not possible;
5. The radioactive waste must be disposed of;
6. An isotope laboratory must be available as well as adequately trained personnel who have regular medical examinations;
7. [^{32}P]-labeled substances are relatively expensive.

[^{3}H]-labeled samples are mostly used for in situ hybridization experiments because the diffusion radiation of this isotope is rather small (Gall and Pardue 1971). This is the prerequisite for generating sharp hybridization ranges. As a result of the low radiation intensity, however, exposition periods of several weeks and often even months are necessary.

Because of the difficulties encountered in the handling of radioactive isotopes, attempts were made in the 1980s to label hybridization probes by nonradioactive means and to detect them with the help of optical systems after its hybridization with complementary target sequences.

For the nonradioactive labeling of hybridization probes, a large variety of different binding principles have been used over the years (Fig. 2). Most important interactions occur between the vitamin biotin and its binding proteins avidin or streptavidin as well as between hapten and antibodies (Langer et al. 1981; Traincard et al. 1983; Herzberg 1984; Porstmann et al. 1985; Hopmann et al. 1987; Hyman et al. 1987; Pezzella et al. 1987; Rashtchian et al. 1987). However, more complex approaches with polynucleotide phosphorylase coupled to bioluminiscence generation have also been applied.

In most cases, the overall reaction is performed in two steps (Fig. 3): (1) Incorporation of a nonradioactive label into nucleic acid, and (2) detection of the nucleic acid-bound label by colour-developing, fluorescence or luminescence systems (Meinkoth and Wahl 1984; Pereira 1986; Höfler 1987; Viscidi and Yolken 1987; Wolf 1987; Donovan et al. 1987; Wilchek and Bayer 1988).

As a first step, the nucleic acid is nonradioactively modified (Fig. 4). High modification rates are achieved by the enzymatic incorporation of modified nucleotides, but chemical or photochemical methods can also be applied here (Urdea et al. 1988). Homogeneous incorporation of the nonradioactive label can be performed enzymatically by nick translation with *E. coli* DNA polymerase I (Langer et al. 1981; Garbutt et al. 1985; Gregersen et al. 1987) or by random-primed synthesis with Klenow enzyme and hapten-labeled nucleotides as substrates (Höltke et al. 1989; Kessler et al. 1989). DNA end-labeling can be obtained by tailing with terminal transferase and modified nucleotides (Riley et al. 1986; Vary et al. 1986).

Single-stranded DNA can be labeled with reverse transcriptase by a primer-catalyzed reaction and de novo synthesis of the complementary strand (Pitcher et al. 1987). In this context, the temperature-stable DNA polymerase from *Thermus aquaticus* (Chien et al. 1976; Kaledin et al. 1980) plays an essential role, since it allows exponential DNA amplification via 20–30 sequential cycles of synthesis using two antiparallel primers (Saiki et al. 1985a, 1985b; Saiki et al. 1986; Mullis et al. 1986; Mullis and Faloona 1988; Oste 1988) and subsequent direct genomic sequencing of the amplified DNA (Wong et al. 1987; Stoflet et al. 1988).

DNA modification ↔ Binding component	Example
Vitamin ↔ Binding protein	Biotin ↔ Streptavidin
Hapten ↔ Antibody	Digoxigenin ↔ Anti-Digoxigenin antibody
Protein A ↔ Constant region of IgG	Protein A ↔ IgG
DNA/RNA hybrid ↔ DNA/RNA-specific antibody	DNA/RNA ↔ Anti-DNA/RNA antibody
RNA/RNA hybrid ↔ RNA/RNA-specific antibody	RNA/RNA ↔ Anti-RNA/RNA antibody
Binding protein DNA sequences ↔ Binding protein	T7 promoter ↔ *E.coli* RNA polymerase
Heavy metal ↔ Sulfhydryl reagents	Hg^{2+} ↔ HS-TNP ↔ Anti-[TNP-DNA] antibody
Polyadenylation ↔ Polynucleotide phosphorylase/ pyruvate kinase	ATP-coupled firefly luciferase reaction

Fig. 2. Tpyes of interaction for nonradioactive DNA detection

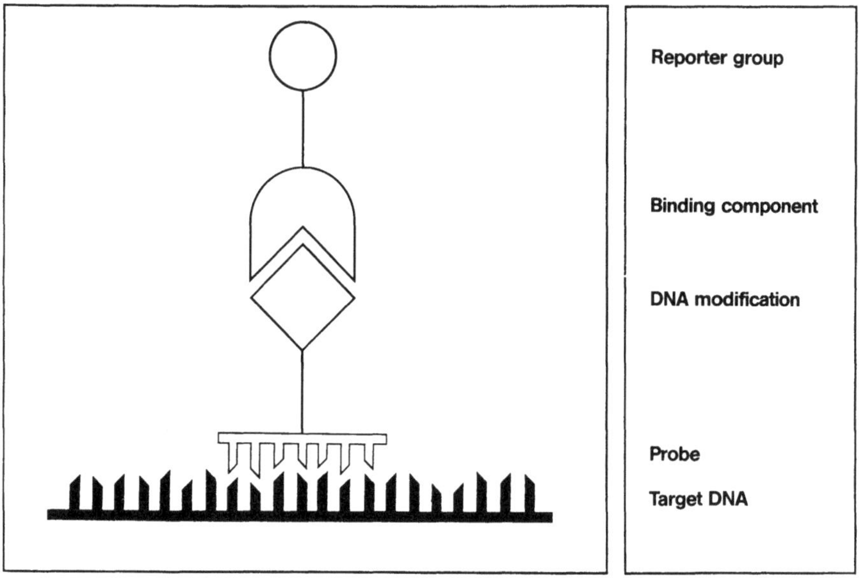

Fig. 3. Principle of noradioactive DNA detection

DNA modifications can also be obtained by photochemical reactions with photoreactive reagents such as azide compounds (e.g. photobiotin: Forster et al. 1985; Lacey and Grant 1987) or nucleic acid-intercalating substances (e.g. psoralen: Brown et al. 1982), as well as by chemical derivation such as sulfite-catalyzed substitution reactions (Viscidi et al. 1986; Gillam and Tener 1986; Reisfeld et al. 1987) or by bromination of nucleotide bases (Keller et al. 1988). Labeling of oligonucleotides can be performed by the insertion of labeled nucleotide derivatives (Cook et al. 1988; Mühlegger et al. 1989).

In a second step, specific detection is performed by binding a color-developing, fluorescence or luminescence system to the modified nucleic acid (Fig. 5; Urdea et al. 1988). In the color-developing systems, labeling enzymes like alkaline

- **Enzymatic nucleotid incorporation**
  - random-primed labeling (Klenow enzyme)
  - nick translation (*E. coli* DNA polymerase I)
  - tailing reaction (terminale transferase)

- **Chemical derivatization**
  - bisulfide catalyzed transamination
  - oligonucleotide synthesis with allylamine-substituted phosphoamidites

- **Light induced modification**
  - azido compounds (photo-derivatives)
  - intercalating psoralen derivatives

- **Binding of basic macromolecules**
  - polyethylenimin/polyethylenglycol poly-enzymes
  - histon H1
  - cytochrome c

- **DNA cross-linking**
  - succinimidyl compounds
  - diepoxyalkyl compounds

Fig. 4. Nonradioactive DNA labeling techniques

phosphatase, β-galactosidase or horseradish peroxidase (Wilson and Nakane 1978; Ishikawa et al. 1983; Inove et al. 1985) are used in combination with suitable color substrates characterized by high extinction coefficients (Horwitz et al. 1966; McGadey 1970; Anderson and Deinard 1974; Franzi and Vidal 1988), which are conjugated to the DNA-binding agents. An alternative system has been described, however, where the marker enzyme itself – in this case alkaline phosphatase – is directly linked to the oligonucleotide probe without additional oligonucleotide modification (Seriwatana et al. 1987; Oprandy et al. 1988). Binding of poly-alkaline phosphatase coupled to basic macromolecules has also been reported (Renz and Kurz 1984; Al-Hakim and Hull 1986; Jablonski et al. 1986; Inoue et al. 1987).

In addition to the color-developing systems, fluorescence (Syvänen et al. 1986; Dahlen et al. 1987; Watson 1987) and luminescence systems (Matthews et al. 1985) are also known. Fluorescence markers like fluorescein (Romanchuk 1982; Voss

- **Chromogenic systems**
  - marker enzyme conjugates
    - alkaline phosphatase
    - horseraddish peroxidase
    - ß-galactosidase
  - oligonucleotide-bound marker enzymes
  - polyethylenimin-coupled poly-enzymes

- **Luminogenic systems**
  - chemiluminiscence
    - horseraddish peroxidase/H_2O_2 system: light amplifiers: luminol, lucigenin
  - bioluminiscence
    - bacterial luciferase system: R·CHO/$FMNH_2/O_2$
    - firefly luciferase system: D-luciferin/ATP/O_2

- **Fluorogenic systems**
  - DNA/protein-bound fluorochrome excitation
    - fluorescein/tetramethyl-rhodamine/phycoerythrin
  - $Europium^{3+}$-ion induced time-resolved fluorometry

- **Metallogenic systems**
  - ferritin coupling
  - gold labeling/silver enhancement
  - magneto-bead coupling

Fig. 5. Nonradioactive DNA detection systems

1984) are often bound directly to the DNA-binding agent and are stimulated by photones of appropriate wavelength (λ_{max} = 490 nm). The best-known luminescence marker is luminol (aminophthalic acid: Puget and Michelson 1976) and the Gundermann reagent (7-dimethylamino-naphthalene-1,2-dicarbonacid-anhydride: Gundermann et al. 1965). These substances are stimulated to emit photons by POD-catalyzed H_2O_2 oxidation which gives rise to oxygen radicals (-O·) and elementary nitrogen (N_2) ($\lambda_{max/Luminol}$ = 430 nm; $\lambda_{max/Gundermann\ reagent}$ = 515 nm).

In addition, bioluminescence systems are also known where, for example, D-luciferin-O-phosphate is used as substrate (Hauber and Geiger 1987). Release is mediated in this case by antibody-bound alkaline phosphatase. D-luciferin is subsequently detected in a coupled reaction by the luciferin enzyme from firefly *Photinus pyralis* (DeLuca and McElroy 1978) in the presence of energy-supplying

ATP and oxygen as oxidizing agent. The oxidation gives rise to oxiluciferin, AMP, and pyrophosphate by emitting photons (λ_{max} = 562 nm). These photons serve as a measuring signal (Wulff 1983).

Nucleic acids can also be made visible with the help of gold particles coupled to hapten-binding antibodies. The gold particles form nuclei to bind a coating of silver layers, similar to photo developers. The silver-enhanced particles can be detected with a light microscope (Saman 1986; Theveny and Revet 1987; Tomlinson et al. 1988).

Many of the systems described are either very prone to disturbances or they lack the necessary sensitivity required for single-copy gene detection of eukaryotic cells.

The best-known, and so far most sensitive, nonradioactive nucleic acid detection system is the biotin:streptavidin system which allows detection in the required subpicogram range (Langer et al. 1981; Langer-Safer et al. 1982; Manuelidis et al. 1982; Singer and Ward 1982; Brigati et al. 1983; Leary et al. 1983). In this system, the vitamin biotin is incorporated as label into the nucleic acid; biotin detection originally took place by the binding of avidin from egg white. Avidin was replaced later by streptavidin, a biotin-binding protein isolated from the *Streptomyces avidinii* bacterium (Green 1975).

An advantage of the biotin:streptavidin system is the binding constant for the biotin: streptavidin interaction, which is extremely high (10^{15} Mol^{-1}). Furthermore, the streptavidin protein has four binding sites for the vitamin biotin of low molecular weight (Green 1975).

Even though nucleic acid detection sensitivity is very high after incorporation of biotin, the biotin:streptavidin system frequently encounters problems regarding elevated background or unspecific side reactions. As a result of the elevated background, the signal/noise ratio is often substantially impaired; the corresponding low signals perish in the background when small nucleic acid quantities are to be detected. Sensitivity is thus reduced.

The elevated background is first of all the result of the unspecific binding of streptavidin-enzyme conjugates to the matrix, or the coupled color-developing system to the solid carrier, such as nitrocellulose or nylon membranes (Kessler et al. 1989). Unspecific matrix absorption is particularly marked on nylon membranes such as nytran, which are more stable than nitrocellulose and thus easier to handle.

Aside from the unspecific matrix binding, another decisive disadvantage of the biotin:streptavidin system lies in the mediation of nucleic acid modification by the endogenous vitamin H (Lardy and Peanasky 1953). The vitamin biotin is ubiquitous within the cell and plays a central role in metabolic processes; the fatty acid synthesis cycle should be mentioned in this context. Substantial problems regarding specificity and background are therefore frequently encountered in the detection of nucleic acids in biological materials with biotinylated probes by the unspecific detection of free endogenous vitamin biotin, e.g. during in situ hybridization. As a result thereof, it is frequently not the nucleus, but mostly unspecifically the cytosol, which is stained by free endogenous biotin during in situ hybridizations (Heiles et al. 1988).

These difficulties are inherent disadvantages of the biotin:streptavidin system, which cannot be eliminated by improving the individual reaction steps. It was therefore of great interest to establish a nonradioactive nucleic acid detection system, at least equally sensitive, which does not have this inherent disadvantage. This was achieved by substituting the biotin:streptavidin interaction by an alternative mode of interaction, which is far less prone to disturbances.

2 Highly Sensitive DNA Labeling and Detection System Based on Digoxigenin:Anti-Digoxigenin ELISA Technique

This novel nonradioactive DNA labeling and detection system is based on the detection of nucleic acid-bound digoxigenin, a cardenolid-steroid, which occurs exclusively in *Digitalis* plants (Hegnauer 1971). Thus the new detection system does not allow unspecific reactions with endogenous cellular substances in other biological materials, a decisive advantage over the biotin:streptavidin system.

Digoxigenin is obtained by the cleavage of sugar residues from the desacetyllanatoside C isolated from a plant extract (Fig. 6). Beside purpureaglycoside A and B, desacetyllanatoside C is a member of the therapeutically highly effective *Digitalis* glycosides from *Digitalis lanata* or *Digitalis purpurea* (Reichstein and Weiss 1962).

The primary glycoside desacetyllanatoside C consists of one molecule glucose, three molecules digitoxose and the steroid-aglycon. Glucose removal in the *Digitalis* leaves gives rise to the secondary glycoside digoxin, additional removal of the three digitoxose residues by acid hydrolysis leads to the aglycon digoxigenin (Reichstein 1962).

Removal of glucose from the genuine glycoside can also be performed with a specific glycosidase isolated from fresh *Digitalis* leaves – the digipurpidase (Neumüller 1981).

Chemically, the cardenolide-steroid digoxigenin is derived from cyclopentanon-perhydrophenantren ($C_{23}H_{34}O_5$; Pataki et al. 1953). With 23 C-atoms, the 5β-cardenolides are structurally similar to the gallic acids; the only difference is that the tertiary OH-group at the C-atom 14 shows β-conformation. Another characteristic is the occurrence of the α,β-unsaturated lactone ring on the C-atom 17. Rings A/B are linked in cis-position, rings B/C in trans-position, and rings C/D again in cis-position. Digoxigenin differs from the closely related digitoxigenin only in the OH-group on C-atom 12; gitoxigenin has this OH-group on C-atom 16 (Beyer 1973).

The new nonradioactive DNA labeling and detection system works according to the ELISA principle: Enzyme-Linked Immuno-Sorbent Assay (Heiles et al. 1988; Höltke et al. 1989; Kessler et al. 1989; Martin et al. 1988; Mühlegger et al. 1989; Schäfer et al. 1988b; Seibl et al. 1989). The overall reaction can be divided into three steps (Fig. 7):

Structure of Cardenolids

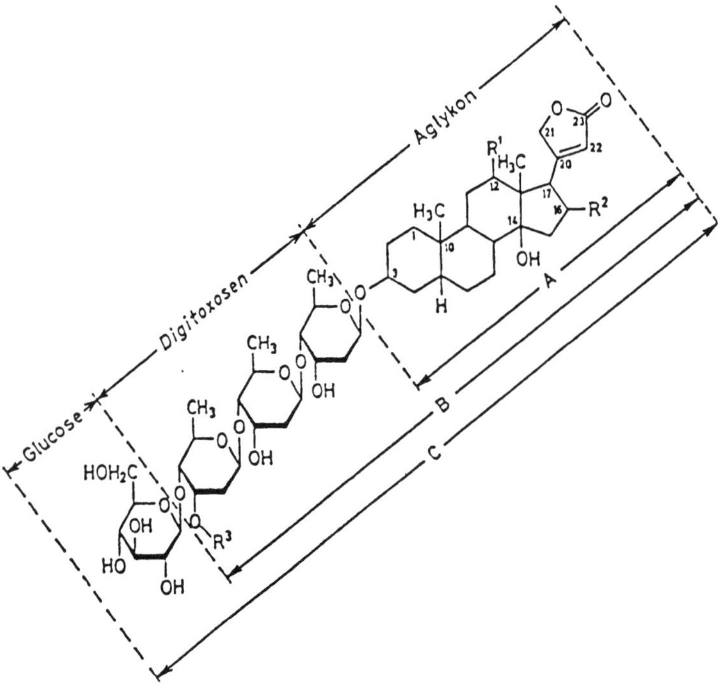

Compound	Part	C_n	Name	R^1	R^2	R^3
Aglycon	**A**	C_{23}	**Digitoxigenin**	H	H	-
			Gitoxigenin	H	OH	-
			Digoxigenin	OH	H	-
Secundary glycoside	**B**	C_{41}	Digitoxin	H	H	H
			Gitoxin	H	OH	H
			Digoxin	OH	H	H
Primary glycoside	C	C_{47}	**Purpureaglykosid A**	H	H	H
			Purpureaglykosid B	H	OH	H
			Desacetyllanatosid C	OH	H	H
		C_{49}	Lanatosid A	H	H	H_3C-CO
			Lanatosid B	H	OH	H_3C-CO
			Lanatosid C	OH	H	H_3C-CO

Fig. 6. Cardenolids of *Digitalis* plants

Preparation of Digoxigenin

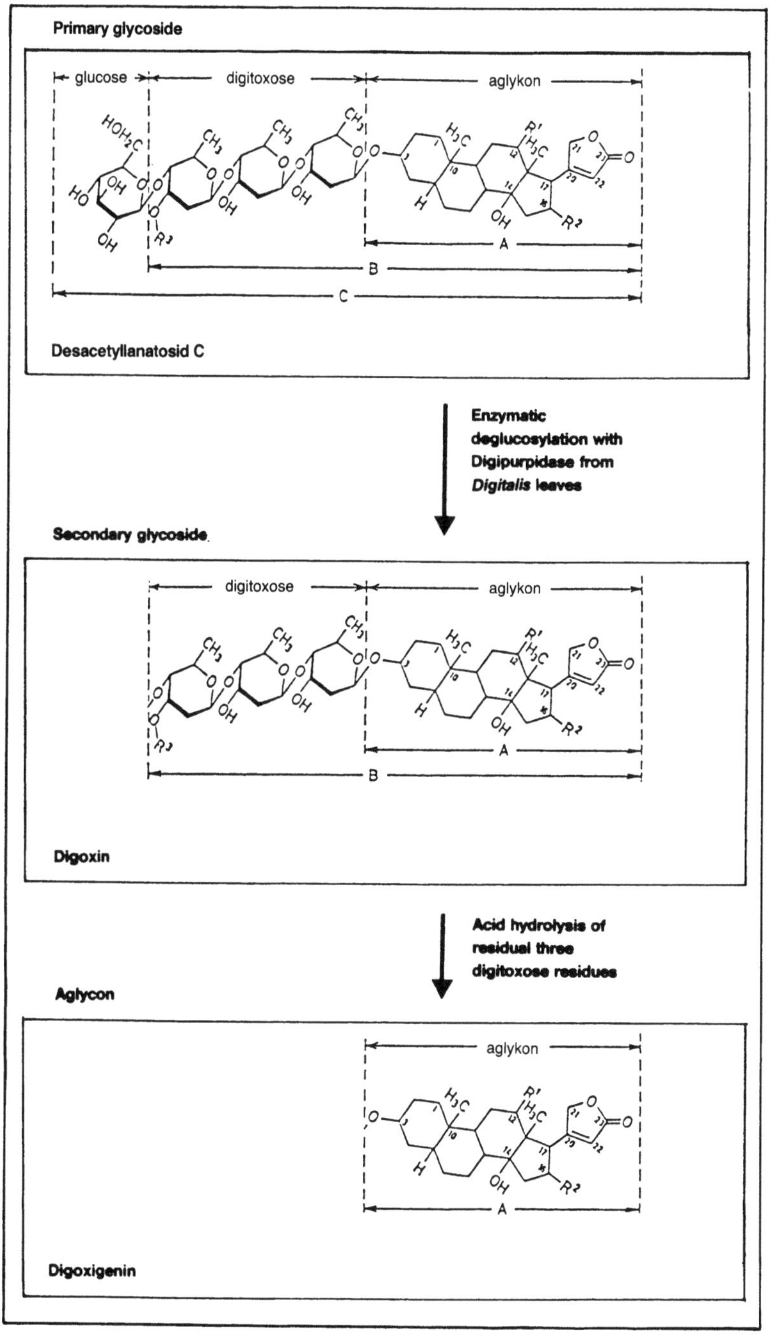

Fig. 6, Part II

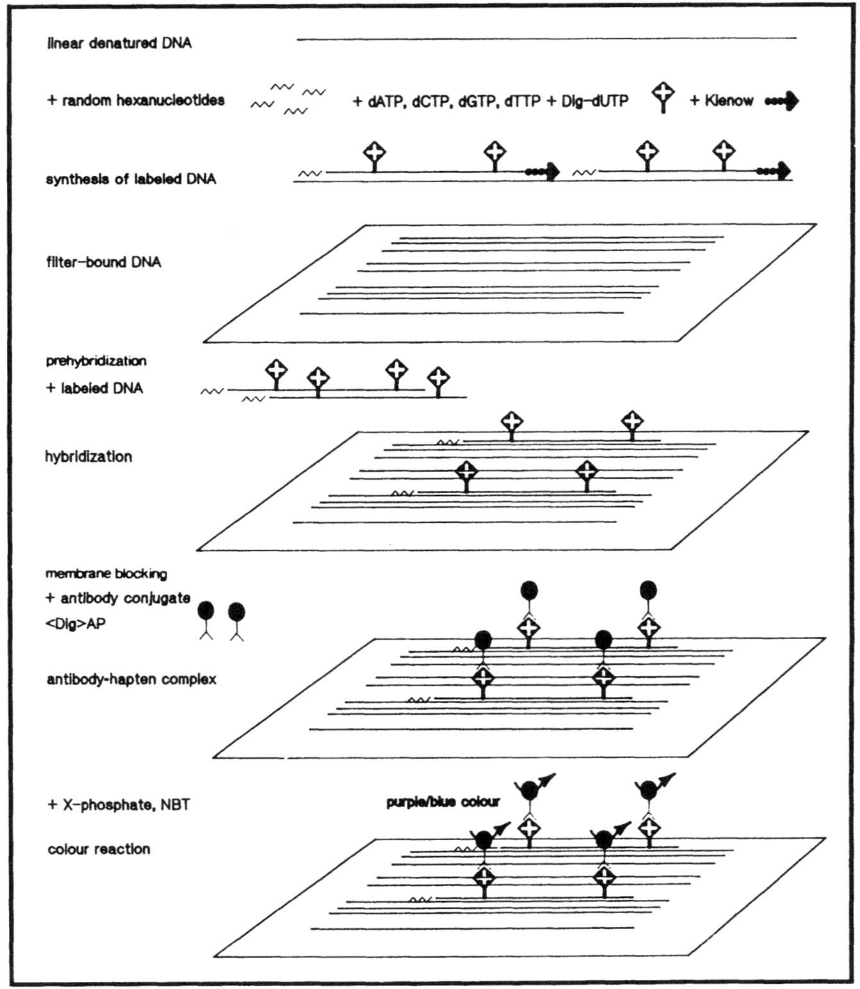

Fig. 7. Nonradioactive DNA labeling and detection with digoxigenin

1. Enzymatic digoxigenin labeling of the DNA probe by incorporation of digoxigenin-labeled desoxyribonucleoside triphosphate(s), like digoxigenin-[11]-dUTP, under random-primed labeling conditions, using Klenow enzyme. The steroid-hapten digoxigenin is linked with the 5C-position of the uracil base via an 11-atom spacer (Fig. 8). Labeling the DNA may also be mediated by the light-catalyzed incorporation of photodigoxigenin (Mühlegger et al. 1989).

2. The sequence-specific hybridization reaction between membrane-bound or otherwise fixed target DNA and the digoxigenin-labeled probe. Hybridization conditions depend on the length and base composition of the hybridizing region. Specific hybrid formation is obtained under stringent conditions (Meinkoth and Wahl 1984; Anderson and Young 1985). Stringency of the hybridization conditions can be controlled by the salt content, pH value, hybridization temperature or the

Digoxigenin – [11] – dUTP

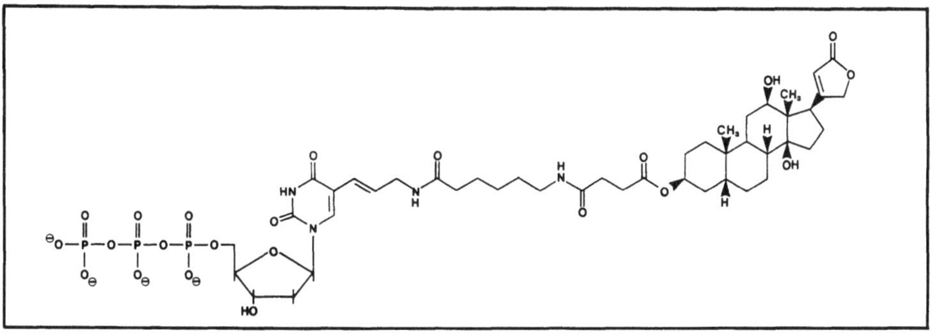

Fig. 8. Digoxigenin-dUTP-derivative for DNA labeling

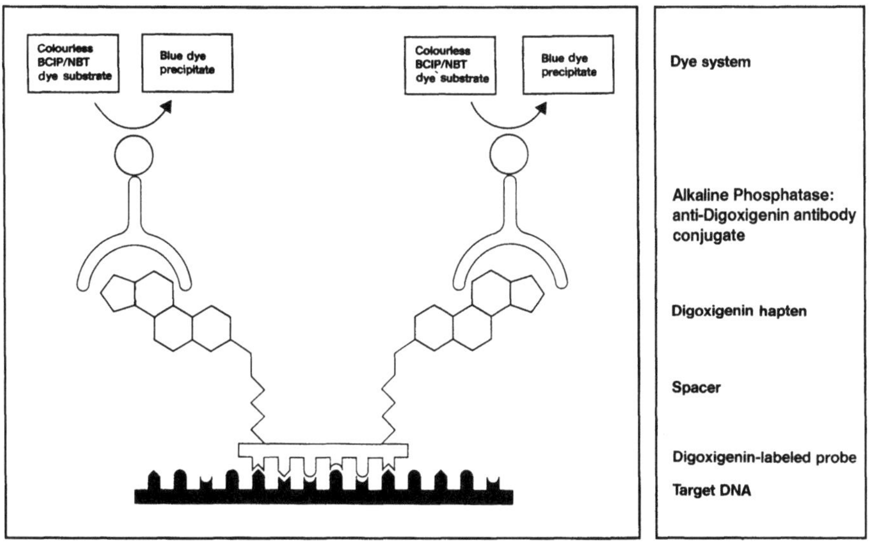

Fig. 9. Nonradioactive detection of DNA based on digoxigenin: anti-digoxigenin ELISA

presence of helix destabilizing agents in the hybridization mixture. Specific hybridization gives rise to immobilized, digoxigenin-labeled hybrid nucleic acid molecules. Nitrocellulose or nylon membranes may be used as carrier.

3. Immunological detection of the digoxigenin-labeled hybrid molecule (Fig. 9). After washing off excessive probe material, the membrane is protected with a blocking agent against unspecific binding reactions. This is followed by high affinity binding of conjugates from digoxigenin-specific antibodies and calf intestine alkaline phosphatase (*CIAP:* Ishikawa et al. 1983) specifically to the DNA-bound digoxigenin haptens (<Dig>:*CIAP*).

The combination of the colorless and soluble reaction educts 5-bromo-4-chloro-3-indolylphosphate (*BCIP*) and the colorless nitro-blue tetrazolium salt (*NBT*) serves as soluble color substrates for the *CIAP*-coupled indicator reaction (Horwitz et al. 1966; McGadey 1970; Anderson and Deinhard 1974; Franzi and Vidal 1988). The reaction products are water-insoluble, deep-blue color precipitates which adhere directly to the membrane (McGadey 1970). Rehybridization on nylon materials is possible after the original color precipitate had been washed off with dimethylformamide (Gebeyehu et al. 1987).

3 Sensitivity, Specificity and Low Background of the Digoxigenin System

In the following section, the high sensitivity and specificity as well as the low background staining of the new system, will first of all be demonstrated with the help of model experiments. Even though the binding constant for the digoxigenin-hapten and digoxigenin-specific antibody interaction ($K = 2\times10^8 - 7\times10^9$ mol^{-1}: Hunter et al. 1982) is several orders of magnitude lower than for the biotin:streptavidin system ($K = 10^{15}$ mol^{-1}: Green 1975), 0.1 pg homologous DNA can be detected in the digoxigenin system in less than 16 h in dot, slot and Southern blots; nitrocellulose or nylon membranes can be used as matrix.

The next figure shows the comparison of sensitivity between digoxigenin-, biotin- and radioactive [^{32}P]-labeling in homologous dot blots (Fig. 10; Kafatos et al. 1979). Linearized pBR328 plasmid DNA served as model DNA. All modifications were performed by an improved random-primed labeling procedure using optimal spacer length: Dig-[11]-dUTP and Bio-[16]-dUTP (Höltke et al. 1989; Langer et al. 1981; Brigati et al. 1983). In each of three analogous dilution series, linearized between 10 and 0.01 pg, denatured pBR328 DNA was spotted. All dilutions, as well as the controls without pBR328 DNA, contained 50 ng herring sperm DNA additionally. After a 1-h prehybridization phase, the buffer was replaced and hybridized for 16 h at 68 °C. The subsequent color reaction and/or autoradiography was performed for an additional 16 h.

In all three cases, 0.1 pg homologous DNA was detected. The control experiment with heterologous herring sperm DNA was negative in all three cases, although 10^5-times higher quantities were applied. This means that, in spite of the

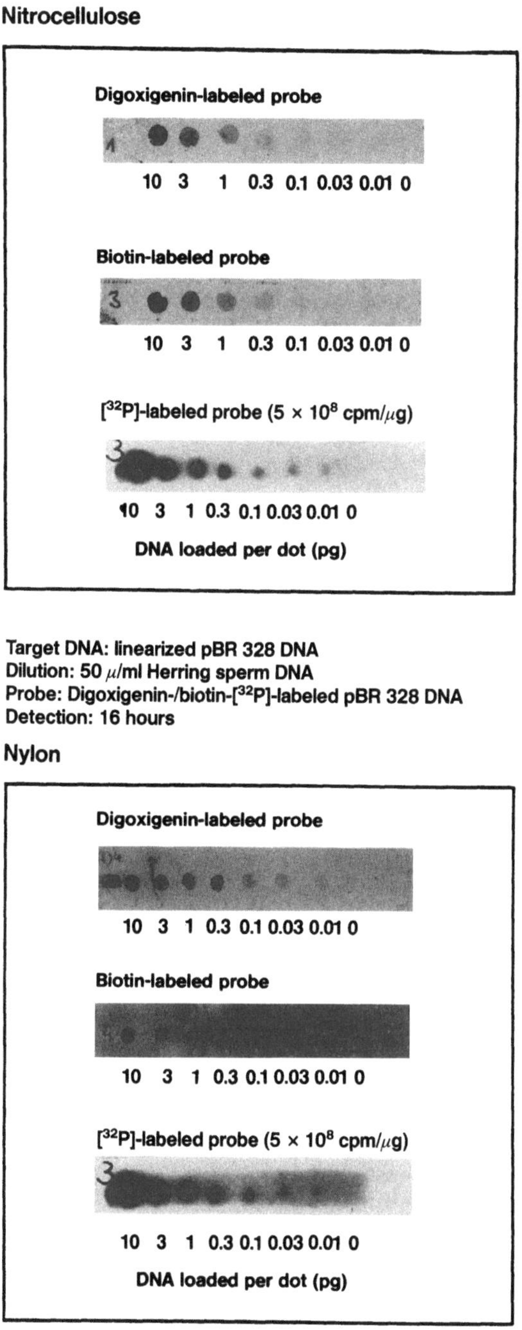

Fig. 10. Comparison of DNA detection in homologous dot blots by hybridization with digoxigenin-, biotin- and [^{32}P]-labeled probes

Digoxigenin-labeled probe

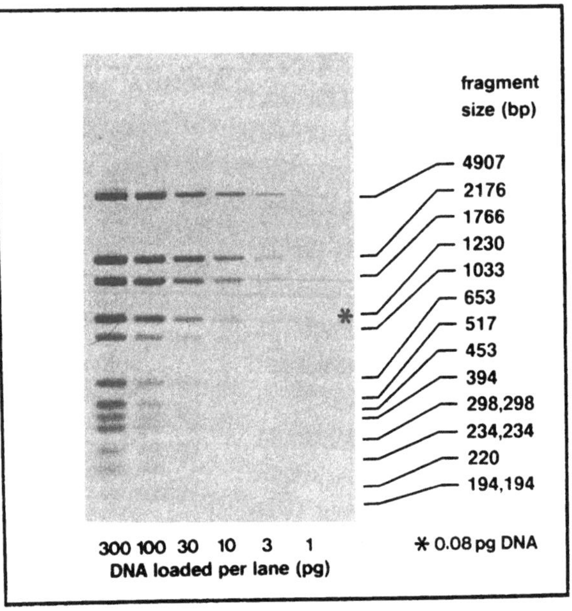

Target DNA: pBR328 was either linearized or digested separately with *Eco*RI, *Bgl*I and *Hin*fI; finally equal amounts were mixed
Dilution: 50 μg/ml Herring sperm DNA
Probe: Digoxigenin-/biotin-labeled pBR328 DNA
Detection 16 hours
Membrane: Nylon

Biotin-labeled probe

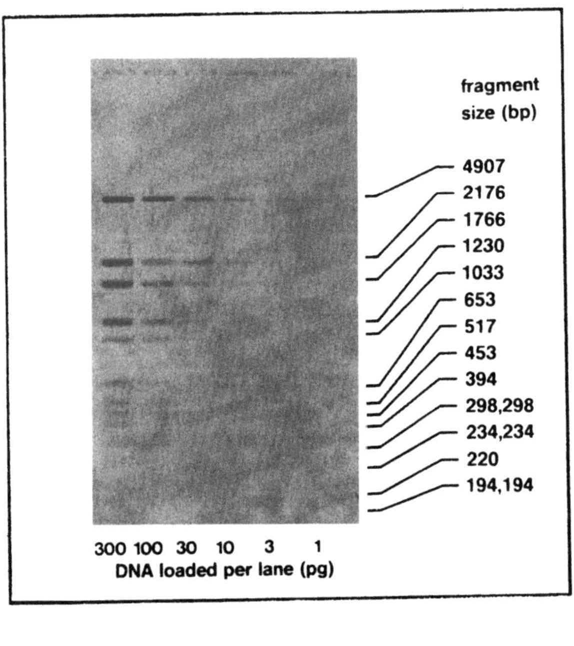

Fig. 11. Sensitivity of DNA detection in homologous Southern blot hybridization

different binding constants in the digoxigenin and biotin system, both nonradioactive detection methods show analogous senstivity and specificity which are comparable with the radioactive [^{32}P]-detection. As regards biotin labeling, however, a distinct background can be observed on the nylon membranes in particular. This significant background occurs in the digoxigenin system neither with nitrocellulose nor with nylon membranes.

In an additional experiment the detection of pBR328 fragments was performed in homologous Southern blots (Fig. 11; Southern 1975). In this experiment, aliquots of pBR328-fragment mixtures between 30 and 1 pg overall DNA were separated by electrophoresis and transferred to nylon membranes. Hybridization was again performed with digoxigenin-labeled pBR328 DNA. With this probe, the 1-pg lane clearly shows the fourth band from the top. The calculated DNA quantity in this band is 0.08 pg. This means that also after Sothern transfer, up to 0.1 homologous DNA can be detected without significant background. The analogous experiment in the biotin:streptavidin system shows similar sensitivity but higher background for this matrix.

In summary, it can be concluded that the new system allows specific detection of 0.1 pg homologous DNA within a detection period of only 16 h. In contrast to the biotin:streptavidin system, significant background occurs neither on nitrocellulose nor on nylon membranes.

4 Optimizing the Individual Steps of the Digoxigenin System

4.1 Labeling

Which are the most important parameters of the three parts of the overall reaction, and which are the optimum reaction conditions of the individual steps?

Firstly, the parameters of the labeling reaction are described. The next figure shows a comparison between enzymatic, photochemical and chemical digoxigenin labeling (Fig. 12; Viscidi et al. 1986; Lacey and Grant 1987; Höltke et al. 1989; Kessler et al. 1989; Mühlegger et al. 1989). As regards the parameter quantitiy of template DNA which can be labeled, the time required for the labeling reaction and the sensitivity which can be attained in the DNA detection reaction, the enzymatic labeling reaction is clearly superior to the other labeling methods.

When comparing nick translation with *E. coli* DNA polymerase I (Rigby et al. 1977) and random-primed synthesis with Klenow enzyme (Feinberg and Vogelstein 1983, 1984), with equal amounts of input template DNA, 1.5- to 2 times more labeled DNA is synthesized with the latter labeling method (Fig. 13). The reaction reaches an optimum after 1–2 h, and after 20 h of incubation only a slight increase can still be observed. In nick translations, even a slight decrease occurs after extended periods of incubation.

In the random-primed method, de novo synthesis of labeled DNA depends on the template quantity used. In the ng range, i.e., with DNA quantities obtained

Method	Amount of template DNA	Reaction time for labeling	Sensitivity in homogeneous dot-blots (detection 16 h)	Advantages	Disadvantages
Enzymatic incorporation of digoxigenin-haptens				● Low amount of input DNA ● High sensitivity	
● Random-primed DNA labeling	0.02 – 3 μg	1h	0.1 – 0.05 pg		
● Nick-translation	0.1 – 1 μg	1h	0.5 – 0.1 pg		
Photodigoxigenin	1 – 50 μg	1h	1 – 0.5 pg		● Equipment ● Darkroom ● High amount of input DNA needed
Chemical incorporation of digoxigenin-haptens	1 – 50 μg	24 – 72 h	10 – 1 pg		● Time ● Reproduction ● High amount of input DNA needed

Fig. 12. Methods of nonradioactive homogeneous labeling of DNA with digoxigenin

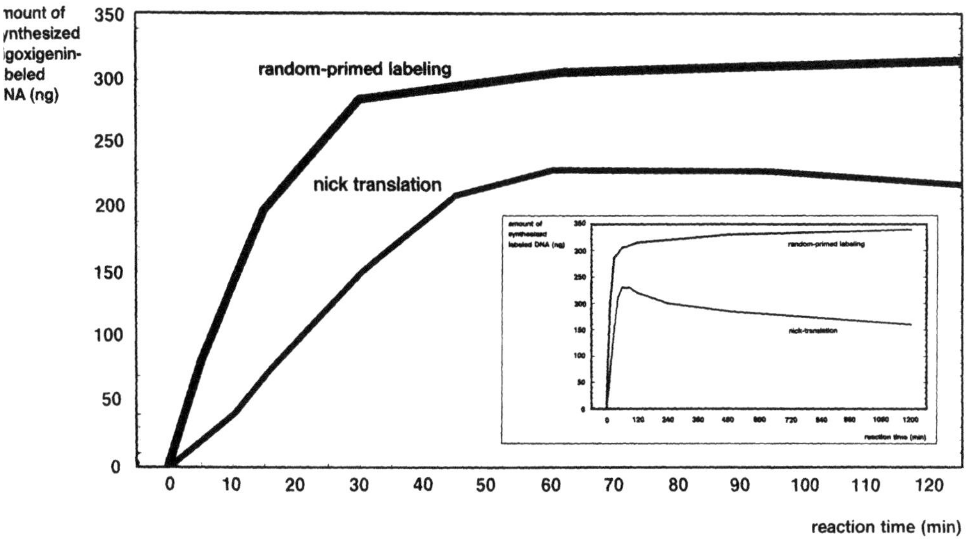

Fig. 13. Kinetics of nonradioactive DNA labeling with random-primed labeling or nick translation

by fragment elution from normally dimensioned agarose gels, de novo synthesis is 30–200%; in the more preparative μg range, the yield is still 18–25%. By increasing the volume of the reaction mixture up to ten-fold, with 1 μg template DNA up to 3 μg digoxigenin-labeled probe DNA is obtained.

The next figure shows the size distribution of the digoxigenin-labeled DNA after labeling under random-primed conditions (Fig. 14). A total of 1 μg pBR328 DNA was labeled. After precipitation with ethanol, the samples were separated by agarose gel electrophoresis with and without prior denaturation, were subsequently transferred onto nylon membranes, and were detected with the anti-digoxigenin:alkaline phosphatase color reaction. This approach allows specific detection only of the newly synthesized DNA. The electrophoresis pattern with the denatured samples shows that during de novo synthesis, a mixture of labeled molecules of different length are formed. The size distribution shows a range between 200 and 2000 bp; the largest quantity of labeled probes is observed in the 300 to 600 bp size range.

Overall sensitivity is decisively influenced by the length of the hydrophilic linear spacer which links the digoxigenin hapten with the 5C-position of the pyrimidine base uracil (Fig. 15; Mühlegger et al. 1989). Best results are obtained with spacer lengths between 11 and 16 spacing atoms; shortening of the spacer leads to losses in sensitivity (Höltke et al. 1989). The reason for this decrease in sensitivity should first of all be caused by the sterical inhibition of the conjugate binding, and secondly, by the disturbance of hybridization due to the hapten being too close to the region of hydrogen bonding.

Aside from the previously mentioned parameters, de novo synthesis of digoxigenin-labeled DNA also depends on the amount of digoxigenin-labeled nucleotides used for labeling (Fig. 16). For this purpose, the ratio between digoxi-

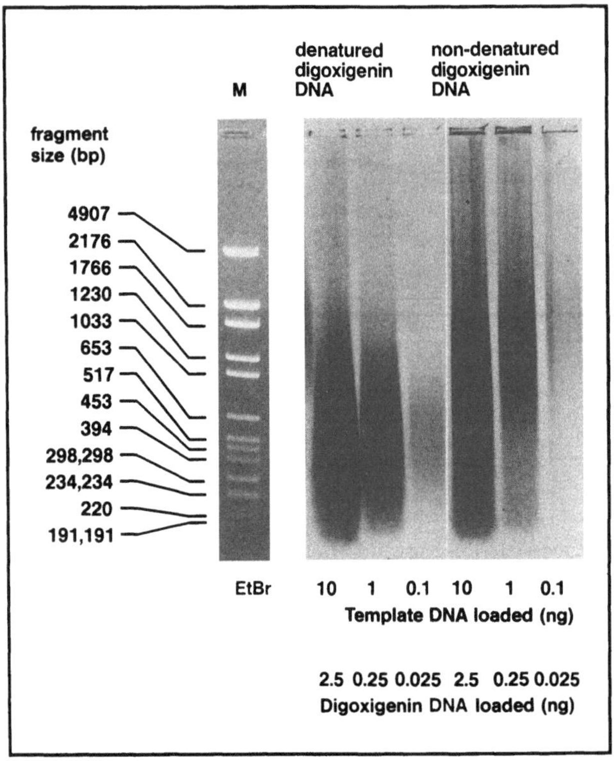

Reaction: Random-primed labeling, 60 min, 37 °C
Template: 1 μg linear pBR328 DNA
M: Molecular weight marker, EtBr stained

Fig. 14. Size distribution of digoxigenin-labeled DNA

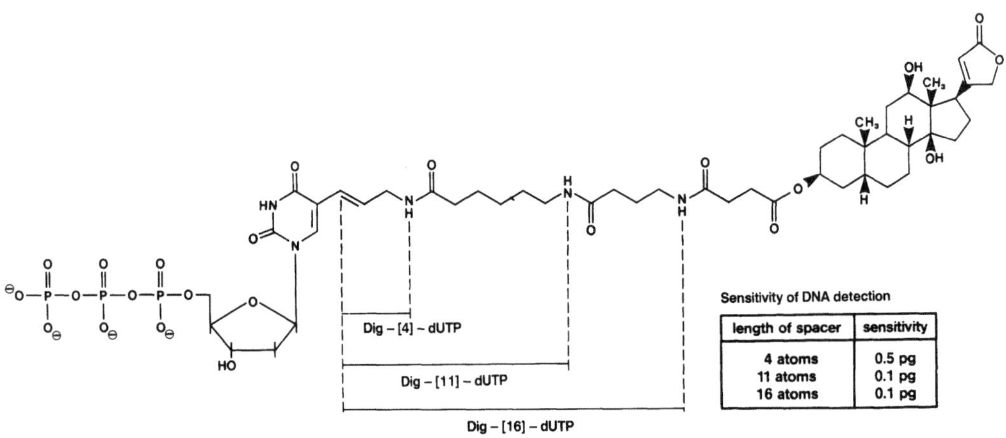

Sensitivity of DNA detection

length of spacer	sensitivity
4 atoms	0.5 pg
11 atoms	0.1 pg
16 atoms	0.1 pg

Fig. 15. Digoxigenin-labeled dUTP with different spacer lengths

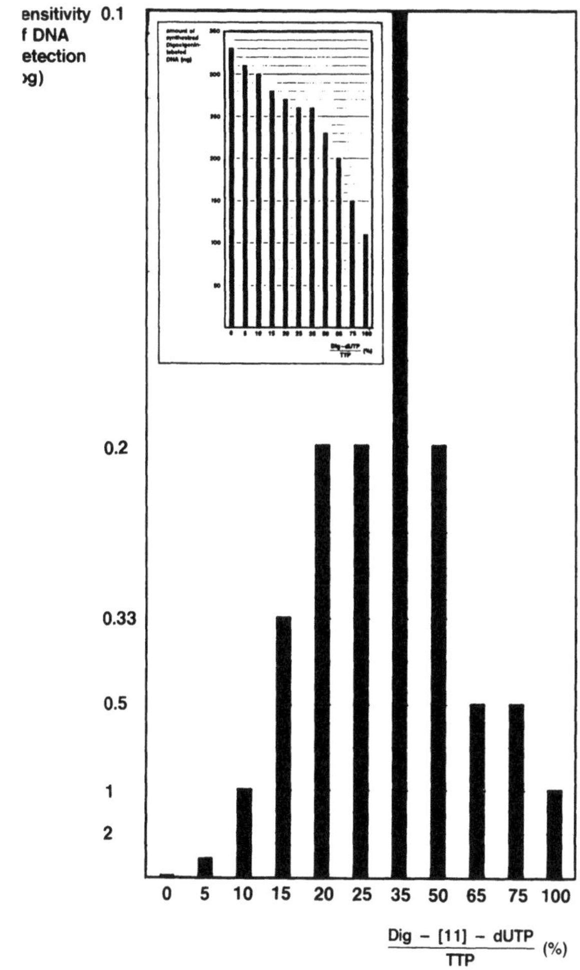

Fig. 16. Effect of ratio of Dig-[11]-dUTP to dTTP in labeling reaction on sensitive of DNA detection

genin-labeled dUTP and unlabeled dTTP was varied. Measurement of the de novo synthesis in this experiment was performed by addition of a constant small [^{3}H]-dUTP quantity as tracer and subsequent TCA precipitation. The concentration dependency shows that with increasing digoxigenin labeling, de novo synthesis of labeled DNA is steadily being reduced. The rate of synthesis obtained with only digoxigenin-dUTP in the absence of dTTP is only approximately 30% of the optimum rate of synthesis obtained in the presence of only unlabeled dTTP.

The sensitivity of the overall system is decisively influenced by the labeling density of the sample. Best sensitivity results in the detection reaction are *not* obtained with the sample showing the highest labeling density, but with the sample labeled with a Dig-[11]-dUTP/dTTP ratio of 33% in the substrate mixture. The labeling density is a critical parameter for the system's overall sensitivity. In a statistical average, only about every 36th nucleotide is labeled under optimum labeling conditions:

1 : 4 (T,A,C,G) × 1 : 3 (Dig-dUTP/dTTP) × 1 : 3
(Dig-[11]-dTUP/dTTP rate of incorporation) = 1 : 36.

Thus, the statistical distance between two digoxigenin-modified nucleotides is about 3.5 helix turns which approximately corresponds to 115 Å or 1.15 μm. The hapten distance shows good correlation with the dimensions of the high molecular antibody:CIAP-conjugates used (Cantor and Schimmel 1980). Figuratively, an excessively high labeling density leads to a sterical inhibition of the conjugate binding. Insufficient hapten labeling, on the other hand, leaves too many free sites on the DNA without the possibility of conjugate binding. Both effects lead to a loss in sensitivity.

Finally, it should be mentioned that the digoxigenin-labeled probe is stable for a minimum of 1 year when stored frozen and that it is directly re-usable after renewed denaturation. This is a distinct advantage over [^{32}P]-labeled probes, since the phosphorous isotope has a half-life of approximately 14 days only.

4.2 Hybridization

The second step of the overall reaction – the hybridization – is possible under all common hybridization conditions (Maniatis et al. 1982; Ausubel et al. 1987). Hybridization can be performed at 68 °C without formamide, but hybridization at 45 °C in the presence of 50 % formamide is equally feasible.

The solutions for pre- and probe hybridization are identical and should only be renewed after prehybridization. The mixture consists of well-established components such as SSC, SDS and sarkosyl, as well as a specially developed blocking

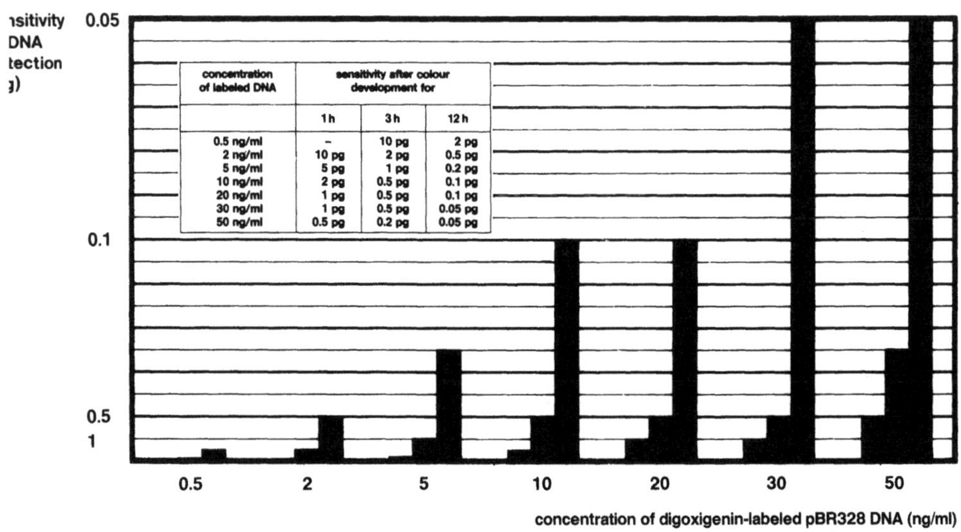

concentration of labeled DNA	sensitivity after colour development for		
	1 h	3 h	12 h
0.5 ng/ml	–	10 pg	2 pg
2 ng/ml	10 pg	2 pg	0.5 pg
5 ng/ml	5 pg	1 pg	0.2 pg
10 ng/ml	2 pg	0.5 pg	0.1 pg
20 ng/ml	1 pg	0.5 pg	0.1 pg
30 ng/ml	1 pg	0.5 pg	0.05 pg
50 ng/ml	0.5 pg	0.2 pg	0.05 pg

Fig. 17. Effect of concentration of digoxigenin-labeled DNA during hybridization on sensitivity of DNA detection

reagent. This reagent was obtained by appropriate fractionation of milk proteins. Adequate volumes of hybridization solution (20 ml per 100 cm^2 membrane) are essential for prehybridization; for probe hybridization 2,5 ml of the same hybridization solution, including the digoxigenin-labeled probe, per 100 cm^2 membrane, are sufficient. The hybridization solution can be re-used, so it is not necessary to overly economize on the use of this solution. The hybridized filters can be used directly in the detection reaction or they can be stored under dry conditions for later use.

Sensitivity and speed of the final detection reaction depend on the concentration of the digoxigenin-labeled probe in the hybridization solution (Fig. 17).

Detection kinetics show that when using 10 to 50 ng/ml digoxigenin-labeled probe in the hybridization solution after addition of conjugate and color substrate, 0.5 to 2 pg homologous DNA can be detected after just 1 h. Sensitivity increases to values between 0.2 and 0.5 pg after 4 h. Full sensitivity of the digoxigenin system is obtained by overnight detection (16 h). Under these conditions, DNA quantities between 0.05 and 0.1 pg can be detected in homologous dot blots. Longer incubation periods, e.g., over a weekend (72 h), do not result in a significant signal increase, even though the background is still low after this prolonged detection period. Detection periods of 16 h and more are not critical; the system is flexible with respect to time.

4.3 Detection

Which are the most important parameters of the final detection reaction? Essential for signal generation above low background is the effective blocking of the membrane against unspecific absorption of the antibody:*CIAP*-conjugate. For membrane blocking, fractionated milk proteins are used as blocking reactions in a concentration of 0.5%, membrane blocking is followed by the addition of high affinity anti-digoxigenin antibodies, which are covalently linked with the alkaline phosphatase marker enzyme to form a high molecular weight complex.

When selecting the marker enzyme, β-galactosidase (β-Gal), horseradish peroxidase (*POD*) and calf intestinal alkaline phosphatase (*CIAP*) were compared (Fig. 18; Wilson and Nakane 1978; Ishikawa et al. 1983; Inove et al. 1985). The most sensitive color developing systems for β-Gal is the chlorophenol red-galactoside (*CPRG*®) substrate, for *POD* the 2,2'-azino-di[3-ethyl-benzthiazolin-sulfonate(6)] (*ABTS*®) substrate and for *CIAP* the 5-bromo-4-chloro-3-indolylphosphate (*BCIP* or *X-phosphate*) substrate in combination with nitro-blue tetrazolium salt (*NBT*) (Horwitz et al. 1966; McGadey 1970; Anderson and Deinhard 1974; Franzi and Vidal 1988). It is essential that a color precipitate with a high extinction coefficient continuously will be formed over a long period of time. This was only true for the *CIAP/BCIP/NBT* system but not for the *POD/ABTS*® system (product inhibition by a dead-end complex after 15 to 60 min) or the β-Gal/CPRG system yielding a soluble color product which has to be immobilized with gelatine or agarose matrices.

Enzyme	Colour substrate(s)	Advantages/Disadvantages
• *CIAP* – alkaline phosphatase from calf intestine – MW = 100 kD – SA = 2500 units/mg	• *BCIP/NBT* – 5-bromo-4-chloro-3-indolylphosphate/ nitroblue tetrazolium salt ➧ blue precipitate	• highly active conjugates • low interference with endogenic material • coupled redox colour system
• *POD* – peroxidase from horseraddish – MW = 40 kD – SA = 1000 units/mg	• *ABTS®* – 2,2'-azino-di-[3-ethylbenz-thiazoline sulfonate (6)] ➧ green precipitate	• active conjugates • interference with endogenic peroxidases • product inhibition
• *β-Gal* – ß-galactosidase from *E. coli* – MW = 540 kD SA = 600 units/mg	• *CPRG®* – chlorophenol red-ß-D-galactopyranoside ➧ red soluble colour product	• active conjugates • low interference with endogenic material • steric hindrance by high MW of *β-Gal* • at present no efficient insoluble dye system

Fig. 18. Commonly used marker enzymes

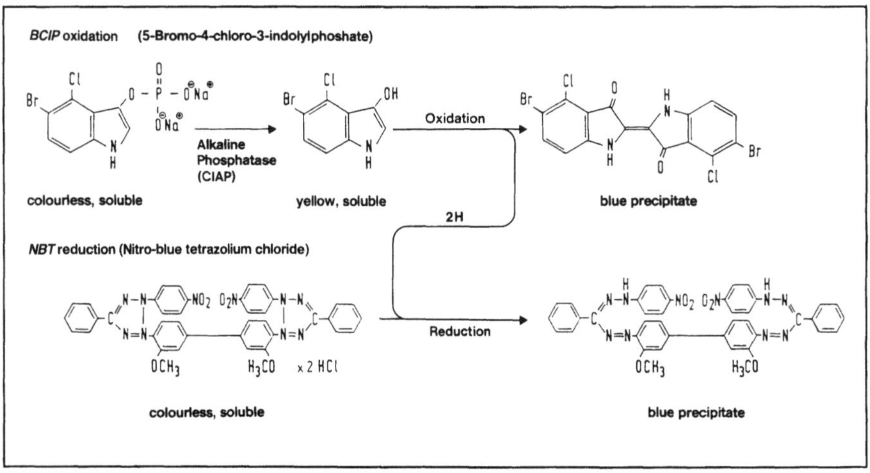

Fig. 19. Substrates and reaction products of alkaline phosphatase (*BCIP*) catalyzed colour reaction with *BCIP/NBT*

With the *CIAP/BCIP/NBT* system in alkaline pH, a deep-blue precipitate is formed from the soluble colorless reaction educts in just a few minutes by phosphate removal – resulting in a yellow soluble intermediate – and a coupled redox-reaction between the generated 5-bromo-4-chloro-3-indolylhydroxyd and *NBT* (Fig. 19). The colored reaction product adheres directly on the membrane and the intensity of the color precipitate increases steadily throughout the entire detection period. Color saturation is reached after 12 – 16 h. On nylon membranes, the color of the precipitate is slightly brownish.

The two color components of the *CIAP/BCIP/NBT* system form a redox system. It is convenient in this case that both the oxidation of the indolyl compound to indigo after the release of the phosphoryl group, and the coupled reduction of the *NBT* to diformazane, lead to the formation of a blue insoluble co-precipitate.

5 Application of the Digoxigenin System in Various Techniques Used in Molecular Biology and Medicine

Aside from the already mentioned dot and Southern blots, the nonradioactive system has so far been applied in the following techniques: genomic blots; colony hybridization; plaque hybridization; virus detection in serum samples (slot blots); in situ hybridization of metaphase chromosomes; in situ hybridization of fixed cells and tissue streaks.

5.1 Genomic Blots: Detection of the Human t-PA Single-Copy Gene

One important aim of the development was a system for the specific nonradioactive detection of single-copy genes in genomic blots. The gene for the human tissue-type plasminogen activator (t-PA) was chosen as an example for a single-copy gene. The DNA used was human placenta DNA.

t-PA is a highly specific serin protease which transforms the plasminogen proenzyme into plasmin (Davie et al. 1986). Plasmin, in turn, specifically dissolves the fibrin network of blood clots. The protein has a molecular weight of 72 kD and consists of a series of domains, which are in part similar to other serin proteases or other plasma proteins.

The human t-PA gene is over 20 kB in length and has a marked intron-exon structure (Fig. 20; Pennica et al. 1983; Fisher et al. 1985; Opdenakker et al 1985). There is a total of 13 short coding exons (II – XIV) correlating with the protein domains (Ny et al. 1984), as well as an additional noncoding exon I at the 5'-end. With one exception (exon XIV: 550 bp), the exons are less than 202 bp in length.

The human t-PA gene was localized by segregation analysis and correlation with corresponding Southern blots as single-copy gene on chromosome 8 (Benham et al. 1984). The t-PA model system gives an answer to the following two questions: (1) Is nonradioactive detection of this single-copy gene possible with the help of short genomic probes? (2) Is nonradioactive detection also possible with cDNA

Intron/Exon Structure of Human t-PA Gene

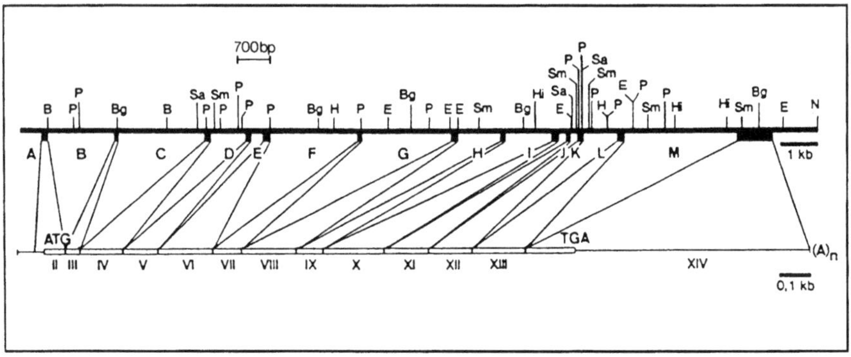

Fig. 20. Physical map of cosmid pcosPAU01 and organization of the human t-PA gene. A restriction map of the cosmid is shown on the top line. The locations of exons II-XIV (solid boxes) and introns A-L were determined by Southern blotting and DNA sequence analysis. The mRNA structure from Pennica et al. (1983) is depicted on the second line, and the open box represents the coding nucleotide sequences. A indicates the poly(A) addition site of the mRNA. The different restriction sites are indicated as follows: B, *Bam*HI; BG, *Bgl*II; E, *Eco*RI; Hi; *Hin*dIII; H, *Hpa*I; N, *Nru*I; P, *Pst*I; Pv, *Pvu*I; Sa, *Sac*I; S, *Sal*I; Sm, *Sma*I (modified after Ny et al. 1984)

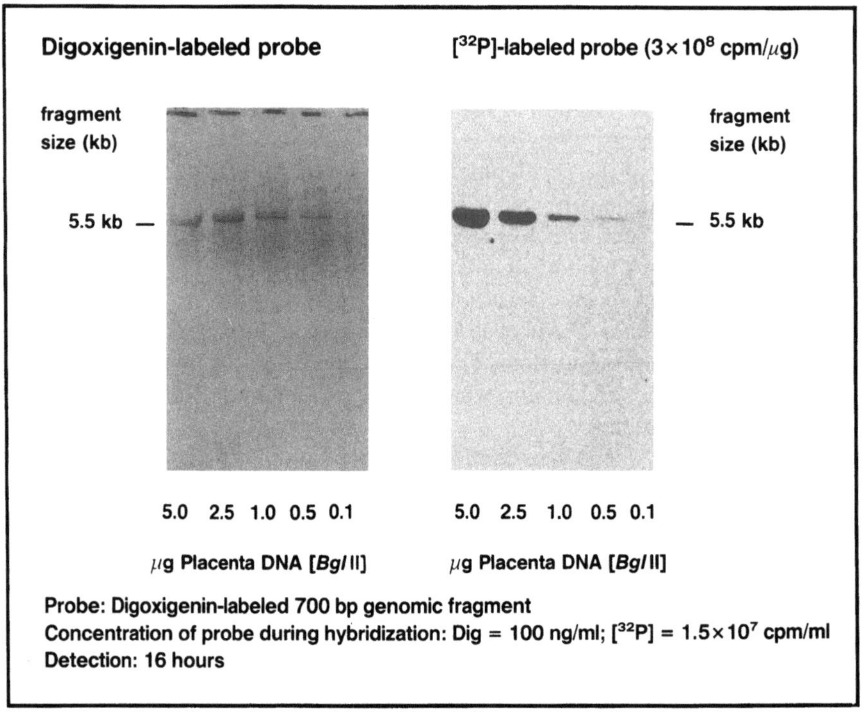

Fig. 21. Detection of the t-PA[*Bgl*II]-fragment in genomic blots of human placenta DNA

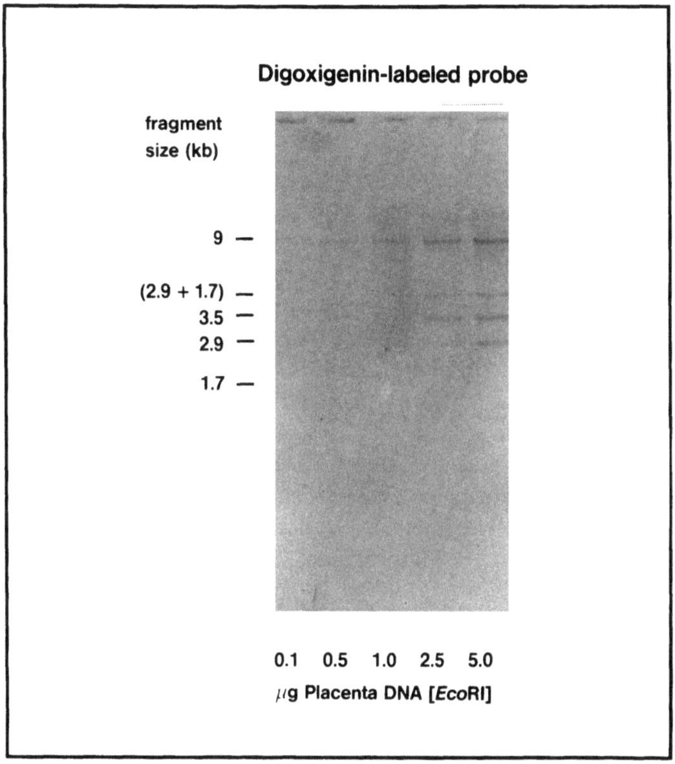

Probe: Digoxigenin – / biotin-labeled cDNA (exons II–XIV)
Concentration of probe during hybridization: Dig = 100 ng/ml;
Detection: 16 hours

Fig. 22. Detection of the t-PA[*Eco*RI]-fragment in a genomic blot of human placenta DNA

probes which only hybridize with the short exon regions? The following experiments demonstrate that both questions can be answered in the affirmative.

In a first experiment, a 700-bp genomic probe (*Pst*I-fragment) was used which covers the region between exon V and VI. In this experiment, the sensitivity of the nonradioactive t-PA detection using a digoxigenin-labeled probe (*Pst*I-fragment) was compared with the analogous detection using a [^{32}P]-labeled probe. For the genomic blot, decreasing quantities of human placenta DNA ranging between 5 and 0.1 μg were cleaved with *Bgl*II and, after gel electrophoresis, were blotted to nylon membranes. *Bgl*II digestion yields a 5.5-kb fragment which contains the region between exon V and VI, and which therefore hybridizes specifically with the applied 700-bp genomic probe (Fig. 21).

Figure 21 shows on the left side the specific detection of this 5.5 kb fragment with the digoxigenin system after application of only 0.5 μg placenta DNA; the detection period was for 16 h. The background is low. The amount of t-PA [*Bgl*II] fragments detected with the 700-bp genomic probe is equivalent to approximately 0.1 pg DNA. Analogous sensitivity is obtained after exposure for 16 h with the

[^{32}P]-labeled sample shown in the middle. The specific activity of this radioactively labeled probe was 5 x 10^8 cpm/μg DNA.

Question 1 can thus be answered as follows: With the digoxigenin system, single-copy gene detection with short genomic probes is possible with a sensitivity analogous to that of radioactive [^{32}P]-detection.

In the second experiment, a cDNA-probe containing all coding exons II-XIV was used (Fig 22). This probe hybridizes only with the short exon regions, not with the intron regions located in between. Decreasing amounts of placenta DNA ranging between 5 and 0.1 μg were cleaved with *Eco*RI and after gel electrophoresis, the fragments were blotted onto nitrocellulose membranes. After hybridization, detection with the digoxigenin-labeled cDNA-probe took again place for 16 h. *Eco*RI-digestion yields 5 t-PA[*Eco*RI] fragments: 9; 3.5; 2.9; 1.7 and 0.1 kb (Fig. 22). In addition, in this experiment a partial band of 4.6 kb – resulting from a polymorphic *Eco*RI site (Benham et al. 1984) – appeared (2.9 plus 1.7 kb). With regard to the different fragments the color intensity is dependent on the size and number of different exons hybridizing to the particular fragments. As in the previous experiment, the large 9-kb fragment can be detected with 0.5 μg placenta DNA. For the detection of the small 1.7 kb fragment 2.5 to 5 μg overall DNA is required.

Question 2 can thus be answered as follows: Sensitive nonradioactive detection is also possible with cDNA probes hybridizing only with short sections of the target DNA. Significant background does not occur on nitrocellulose.

In summary, it can be concluded that with the novel nonradioactive DNA labeling and detection system, single-copy gene detection in 0.5 – 5 μg placenta DNA is possible depending on the nature of the probe and the size and structure of the gene to be detected. This is true both for detection with short genomic probes and for cDNA probes which hybridize discontinuously to genomic DNA fragments.

5.2 Colony Hybridization: Detection of the SUP6-tRNA Gene from Yeast

The following model system was used for testing the use of digoxigenin detection system in colony hybridization (Grunstein and Hogness 1975; Haas and Fleming 1988):

A 750-bp *Bam*HI fragment of the SUP6-tRNA gene from *Saccharomyces cerevisiae* (Wallace et al. 1980) was cloned in pUC19 and transformed in *E. coli* JM109. As a control, the pUC19 vector alone was transformed in *E. coli* JM109. Subsequently, insert-containing bacteria and control transformants were alternatively streaked out on nitrocellulose. After incubation of the bacteria containing membranes on agar plates containing growth medium for 5 h at 37 °C, the *E. coli* cells were treated with alkaline and washed. The released DNA was fixed to the membrane by cross-linking for 3 min at 354 nm (Khandjian 1987). The nonbound bacterial material was removed by another 3-h washing procedure at 50 °C, followed by prehybridization and probe hybridization with digoxigenin-labeled

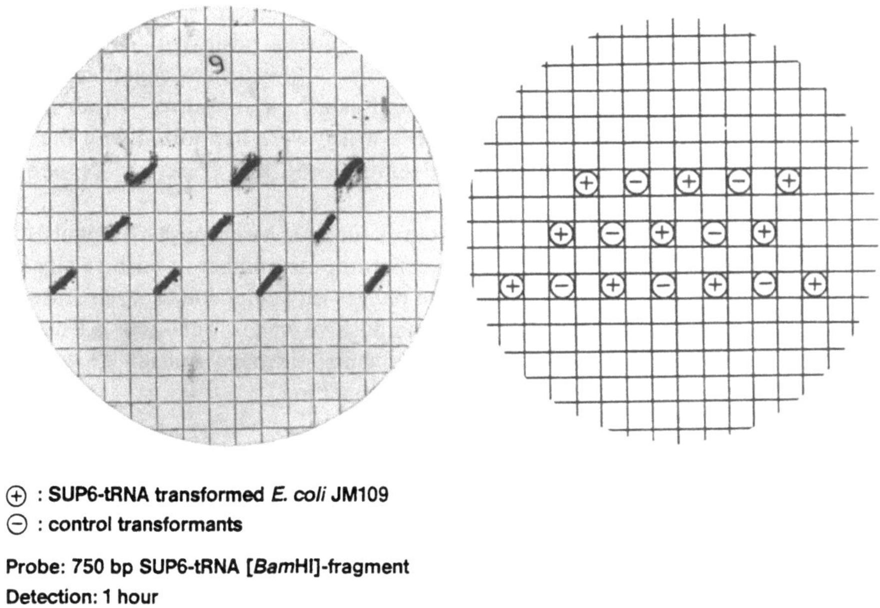

Fig. 23. Colony hybridization: detection of SUP6-tRNA genes from *S.cerevisiae*

SUP6-tRNA[*Bam*HI] fragments. The subsequent detection of recombinant DNA of transformed bacteria was obtained by adding conjugate and color substrates and can easily be completed in less than 1 h.

The result of this experiment is shown in (Fig. 23). After only a short detection period of 30 min, the insert-containing bacteria already give clear signals: the control transformants are not detected. Longer detection periods enhance the positive signals; with the control transformants the background increases only slightly, even after longer detection periods.

5.3 Plaque Hybridization: Detection of the *E. coli* Penicillin G Acylase Gene

As an example for plaque hybridization (Benton and Davis 1978), the detection of the *E.coli* penicillin G acylase gene in λ phages will be demonstrated.

Penicillin G acylase displays – apart from its use as a tool for the production of semisynthetic penicillins – a number of basic properties in prokaryotic cells. Catalytically active enzyme is an $\alpha\beta$ heterodimer and is localized in the periplasmatic space of producing *E.coli* cells (Böck et al. 1983a, 1983b).

For colony hybridization, λgt10 phages containing a *Eco*RI fragment of the *E. coli* penicillin G acylase gene (Schumacher et al. 1986) were incubated on indicator plates. About 500 plaques were obtained. On another plate, a mixture of penicillin G acylase recombinants and an identical amount of phages, consisting of a heterogeneous population from a gene bank, were plated. On this plate, only the penicillin G acylase recombinants were to give a positive signal with the digoxigenin-labeled penicillin G acylase probe. A third plate containing only λgt10 vector molecules without insert served as a further control. Of all three plates, lift-offs were made on nitrocellulose which were hybridized with the digoxigenin-labeled probe containing the entire penicillin G acylase sequence (Fig. 24).

After 30 min of detection, the first two filters showed positive signals for penicillin G acylase recombinants. In the mixture, the number of positive signals corresponds to the number of recombinant phages with penicillin G acylase insert. The λgt10 phages with other inserts give no positive signal. The same is true for the controls which only contained recombinants without penicillin G acylase insert. This demonstrates the specificity of the method.

5.4 Slot-Blot Analysis: Detection of Hepatitis B Virus (HBV) in Human Sera

As an example for a slot-blot analysis, the detection of hepatitis B virus (HBV) sequences in human sera is shown. HBV infections cause inflammable liver diseases (Redeker 1975). The virus has a circular genome of approximately 3.2 kb which is only partially double-stranded (Delius et al. 1983).

For HBV detection on a DNA basis, a digoxigenin-labeled 1.8 kb *Bam*HI fragment (Su et al. 1986) was used as probe which contains the core antigen region as well as parts of the pre-S1, pre-S2 and X antigens (pos. 1392 to pos. 30). Parallel to the digoxigenin detection, [^{32}P]-labeled probes were also used.

The human sera under investigation are classified as positive or negative on the basis of the immunological markers HB_sAg, HB_eAg, anti-HB_s, anti-HB_c, anti-HB_e and anti-HB_c-IgM. Following alkali treatment with subsequent neutralization, the DNA of the different serum samples was collected on nylon filters in a minifold device. Subsequent filter treatment was performed according to the standard protocol, i.e. baking, pre- and probe hybridization, and final overnight detection.

Figure 25 shows on the left side that in the digoxigenin system, signals are only obtained with positive and not with negative sera. As expected, the control reaction with serum of a vaccinated person is also negative. A comparison with the corresponding radioactive [^{32}P]-detection method shows analogous specificity.

Nonradioactive HBV detection is very sensitive. In a dilution series, HBV sequences are detectable up to a serum dilution of 10^{-3}. This corresponds to the sensitivity of the radioactive system. In the nonradioactive system, no signal is obtained with the corresponding negative sera under any dilution conditions, demonstrating once again the specificity of the digoxigenin detection method.

In addition to the characterization of sera by immunological markers, the new nonradioactive system based on the digoxigenin:anti-digoxigenin ELISA principle is suitable for specific and sensitive serum diagnostics on a DNA basis.

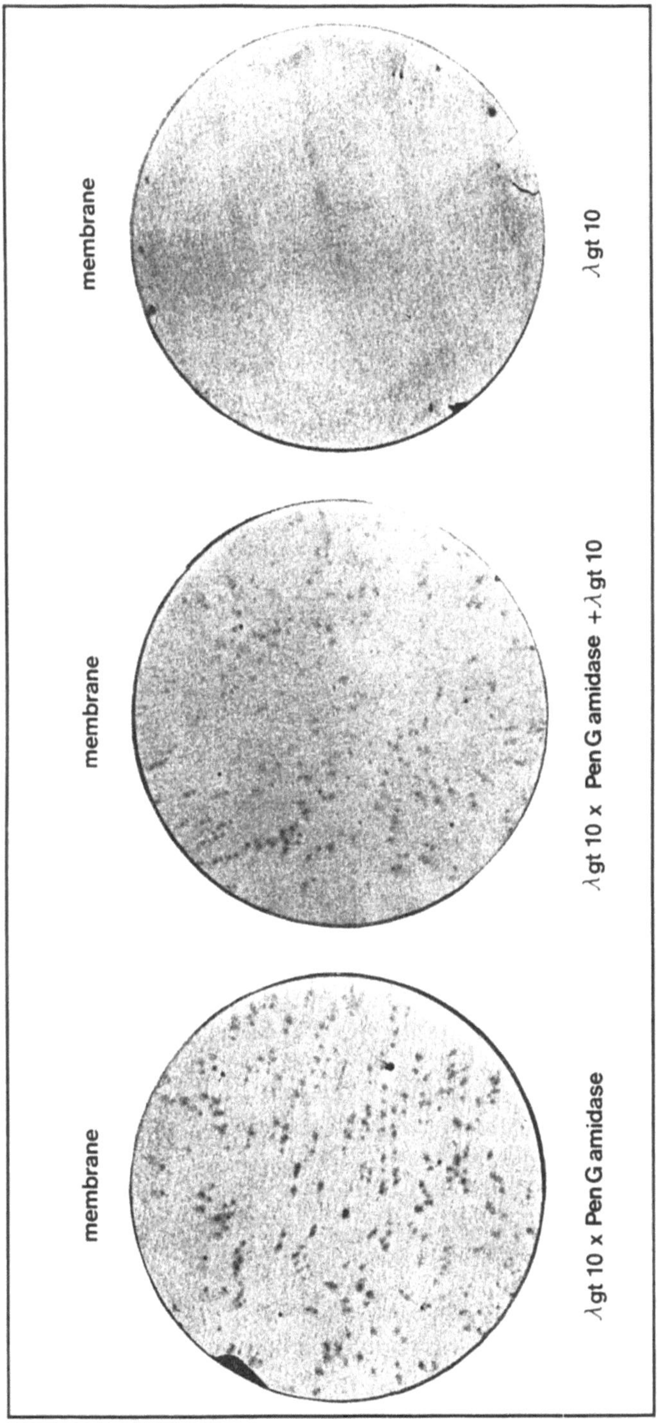

Probe: Digoxigenin-labeled 1.7 kb Pen G amidase [*HpaI*]-fragment
Detection: 1 hour

Fig. 24. Plaque hybridization: Lambda plaques with *E. coli* penicillin G amidase gene inserts

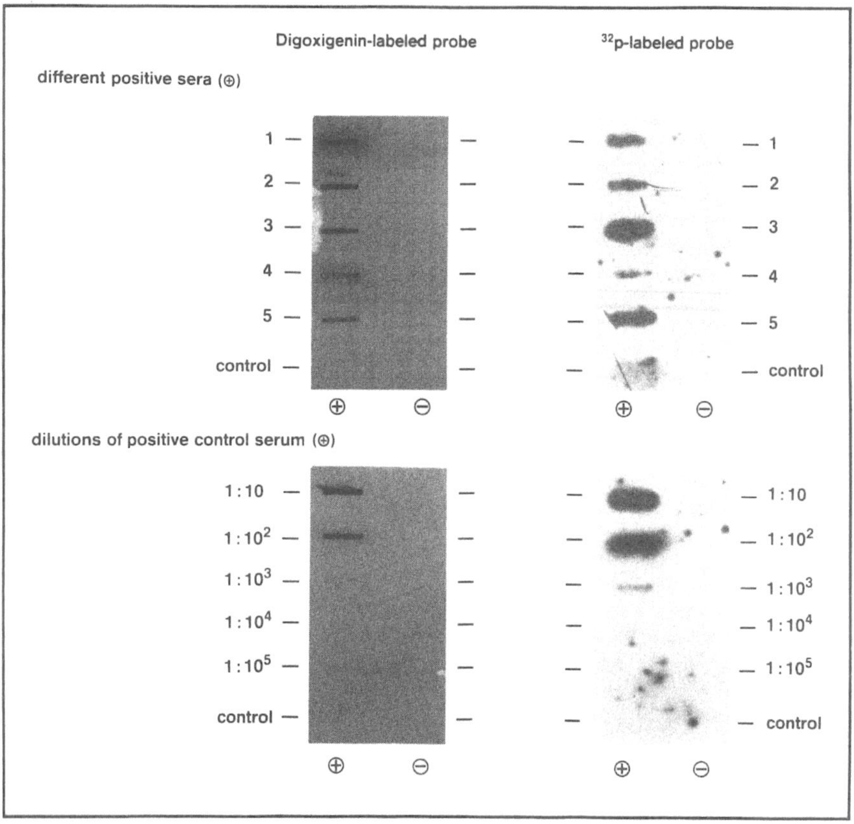

Sample volume: 50μl (positive/negative sera)/10μl (control sera)
Probe: 1873 bp HBV [*Bam*HI]-fragment
Detection: 16 hours

Fig. 25. Detection of hepatitis B virus sequences in human sera (Courtesy of Drs. R. Raßhofer and M. Roggendorf)

5.5 In Situ Hybridization: Detection of t-PA Genes in Metaphase Chromosomes of CHO Cells

So far, in situ hybridization of metaphase chromosomes are mostly performed with [^{3}H]- or [^{35}S]-labeled probes. [^{32}P]-labeled probes are not suitable because of the strong dispersion of radiation. Due to the low radiation intensity of tritium-labeled probes, long detection periods of up to 3 to 4 weeks were so far required. With biotin-labeled probes increased background is observed with signals of low intensity (Ambros et al. 1986).

With the new nonradioactive digoxigenin system, the detection period can be reduced to merely a single night. In addition to being timesaving, the digoxigenin system has an additional advantage – the bands are sharper than those obtained with [^{3}H]-labeled probes. When corresponding fluorescence markers are used,

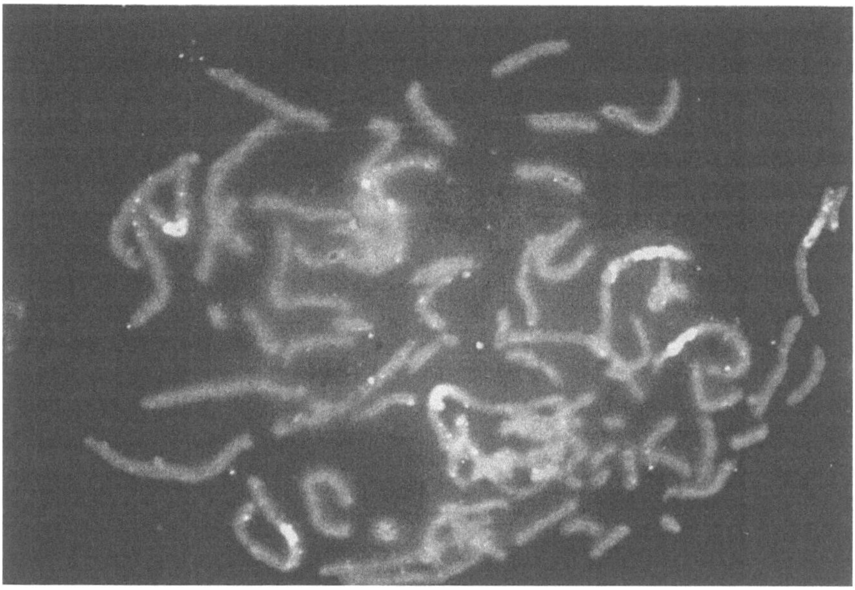

Fig. 26. Detection of cassette-like amplified t-PA genes in metaphase spreads with a digoxigenin-labeled cDNA t-PA probe. Hybridization was visualized by reflection-contrast microscopy. Magnification 100 x (Courtesy of Dr. J. Wienberg)

chromosome banding should additionally be possible. This would only require a change of filter during microscopy.

For in situ hybridization of metaphase chromosomes, CHO cell lines were used which contain t-PA genes amplified in clusters with a copy number of 1000 to 2000. Amplification occurred after transformation of the CHO cells with a vector on which the t-PA gene is coupled with the DHFR region (Weidle et al. 1988). Co-amplification of t-PA and DHFR occurred after methotrexate treatment. The amplified t-PA are arranged like cassettes in homogeneously staining regions (HSR) on the chromosomes.

For the detection of the various t-PA cassettes on metaphase chromosomes, digoxigenin-labeled t-PA-cDNA was used as probe for the in situ hybridization. Fig. 26 shows that after overnight detection, at least five t-PA gene cassettes can be detected by interference reflection microscopy (Verschueren 1985). The results are clearly distinct regions. Each t-PA gene cassette consists on average of 150–300 t-PA genes arranged in tandem repeats.

It is presently under investigation whether t-PA detection is possible in CHO cell lines with smaller t-PA copy numbers, preferably to the level of single-copy t-PA genes.

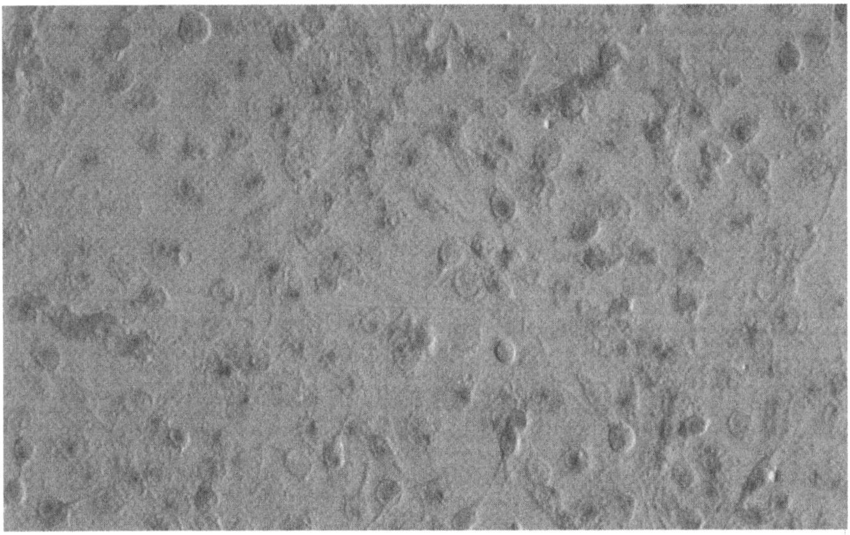

a

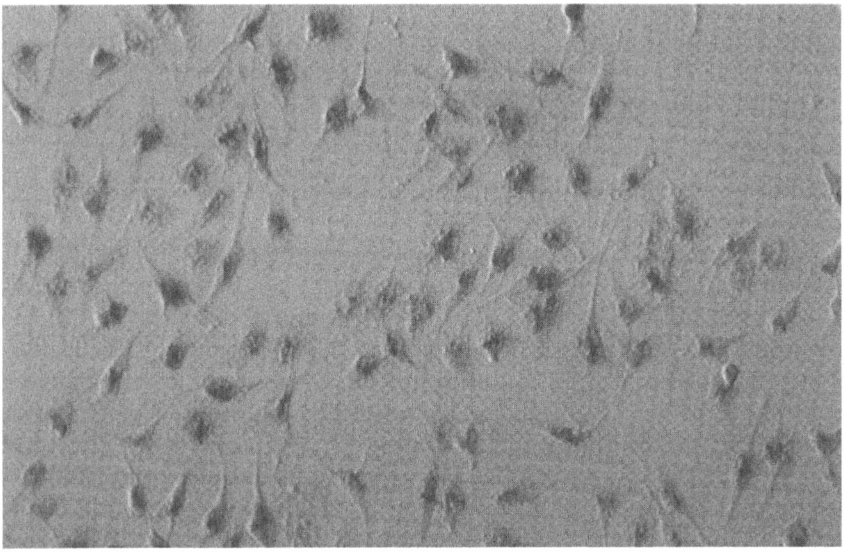

b

Fig. 27. a HeLa cells after hybridization with a digoxigenin-labeled HPV 18 DNA probe purified by vector adsorption. Magnification 340 x (Courtesy of Drs. B.J. Heiles, E. Genersch, R. Neumann, H.J. Eggers). **b** SiHa cells after hybridization with a digoxigenin-labeled HPV 16 DNA probe purified by vector adsorption. Magnification 340 x (Courtesy of Drs. B. J. Heiles, E. Genersch, R. Neumann, H. J. Eggers). **c.** GMK cells after hybridization with a digoxigenin-lalbeled HPV 16 DNA probe purified by vector adsorption. Magnification 340 x (courtesy of Drs. B. J. Heiles, E. Genersch, R. Neumann, H.J. Eggers)

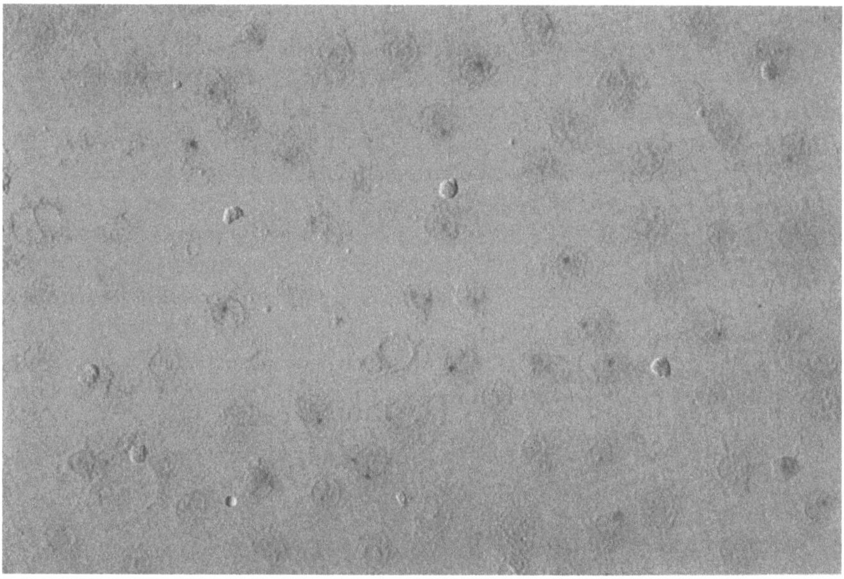

Fig. 27 c

5.6 In Situ Hybridization: Detection of HPV Sequences in Fixed HeLa and SiHa Cell Lines

In situ hybridization (Gall and Pardue 1969; Moench 1987) with digoxigenin-labeled probes was also performed with carrier-fixed smear cells on slides (Heiles et al. 1988). This type of analysis was only reported with the help of [^{35}S]-labeled probes so far (Gupta et al. 1978). In certain cell lines, specific sequences of the human papillom virus (HPV) were detected.

Compared to radioactive labeling, nonradioactive in situ hybridization with smear cells on slides has the advantage of shorter detection periods. The cells were grown directly on the slides. After acid treatment and proteinase K digestion the lysed cells were fixed with paraformaldehyde. After further washing and fixing the cells, DNA was denatured directly on the slides in the presence of the digoxigenin-labeled probe and, after sealing with rubber gum, hybridized under stringent conditions. Following washing and blocking, the detection reaction occurred for 16 h.

For nonradioactive HPV detection, two cell lines were used: (1) the HeLa cell line containing 40 to 200 copies HPV, subtype 18 (Popescu et al. 1987), and (2) the SiHa cell line containing 1 to 2 copies HPV, subtype 16 (Schwarz et al. 1985; Baker et al. 1987; Mincheva et al. 1987).

A digoxigenin-labeled subfragment of the HPV genome, subtype 16 or 18, was used as probe. For hybridization, 20-ng labeled samples per hybridizing area were used. The following figures show HPV detection in HeLa cells in the digoxigenin system (Fig. 27a). The staining of the nuclear region is clearly visible.

DNase digestion before hybridization only left a poor unspecific background; commission of the labeled probe resulted in unstained cells.

Also, in SiHa cells with only one to two copies per cell integrated into the genome, HPV detection was again possible overnight (Fig. 27b); on these cells, staining occurred mainly in the nuclear region. DNase digestion led to poor background; commission of the labeled sample resulted in unstained cells.

HPV-free GMK control cells (green monkey kidney cell line; Eggers 1977) showed neither unspecific staining nor significant background (Fig. 27c).

With in situ hybridizations of fixed cell lines, the new digoxigenin system has the advantage over radioactive systems of being timesaving without significant loss of sensitivity. The unspecific staining often observed with the biotin:streptavidin system (Heiles et al. 1988) is not observed with the digoxigenin:anti-digoxigenin system, due to the rare natural occurrence of the steroid alkaloid digoxigenin.

6 Summary and Perspective

The most important characteristics of the new nonradioactive DNA labeling and detection system are summarized as follows:

1. The new system is based on the ELISA principle and uses the specific interaction between DNA-bound digoxigenin haptens and high affinity anti-digoxigenin antibodies.
2. The new system allows specific detection of 0.1 pg homologous DNA in dot or Southern blot analysis after overnight detection.
3. The high sensitivity of the new system allows single-copy gene detection with 0.5 to 5 μg overall DNA in genomic blots after overnight detection.
4. No significant background occurs in the digoxigenin system in dot, slot, Southern and genomic blots, neither on nitrocellulose nor on nylon membranes.
5. The new system can also be used for colony and plaque hydribizations.
6. The new system also allows specific and sensitive serum analysis on a DNA basis.
7. Furthermore, the new system allows in situ hybridization of metaphase chromosome spreads.
8. Finally, specific and sensitive DNA detection by in situ hybridization is also possible in biological materials such as smear cells on slides.

The new nonradioactive DNA labeling and detection system is intended to provide the basis for the substitution of radioactivity in hybridization techniques which are of increasing importance in basic molecular biology research, in addition to the study of the architecture and expression of pro- and eukaryotic genes (Maclean et al. 1983, Kahl 1988) during construction of recombinant DNA (Maniatis et al. 1982; Glover 1985a, b, 1987; Ausubel et al. 1987), as well as in applied fields of medicine and biotechnology (Yolken 1988).

Non radioactively labeled probes may be applied as DNA markers for the construction of genetic linkage maps of the human chromosomes (White et al. 1985; Donis-Keller et al. 1987; Nakamura et al. 1987). The new system might also be used for genomic sequencing according to Church and Gilbert (1984), as well as for a recently developed sequencing method, where the base-specific bands are blotted on a moving membrane upon leaving the sequencing gel (Beck 1987). Genomic sequencing may be combined with prior DNA amplification so that sequence analysis of small quantities of DNA will be possible (Saiki et al. 1985a, b, 1986; Mullis et al 1986; Wong et al. 1987; Mullis and Faloona 1988; Stoflet et al. 1988). An analogous approach is reported for RNA amplification with transcript sequencing (Sarkar and Sommer 1988). In addition, the new system may be used for criminalistic investigations in forensic medicine, or paternity testing by DNA fingerprinting on a nonradioactive basis, via extensive restriction fragment length polymorphism associated with moderately repeated hypervariable minisatellite sequences (Wyman and White 1980; Jeffreys et al. 1985a, b; Gill et al. 1985; Vassart et al. 1987; Ali et al. 1986; Dykes 1988; Schäfer et al. 1988 a, b, c).

Detection of hypervariable minisatellites by nonradioactively-labeled DNA probes are also useful in domestic animal breeding (Georges et al. 1988). Furthermore, plants which often show high levels of endogenous biotin (Epanchinov 1976; Bilai and Shcherbina 1977; Watanabe et al. 1982) can be studied for the presence or absence of specific resistance genes (Vaeck et al. 1987).

Also, during spreading of viral or bacterial biocides like *Bacillus thuringiensis,* propagation control via nonradioactive DNA detection systems is conceivable (Kamdar and Jayaraman 1983; Holck and Meek 1987; Craviato et al. 1988; Shimizu et al. 1988). Finally, viroid or other subviral pathogen infections of plant seedlings or large plantage areas can be prevented by systematic control with nonradioactively-labeled DNA probes (Gross et al. 1982; Keese and Symons 1985; Riesner and Gross 1985; Gross 1987).

The new nonradioactive system is designed to substantially facilitate analysis of biological material for the presence or absence of certain DNA sequences (Kessler et al. 1989). Therefore, it may be predominantly applied in medical research. Early and easy diagnosis of viral (Kandolf et al. 1987; Ou et al. 1988; Richman et al. 1987; Swierkosz et al. 1987; Ticehurst et al. 1987) and bacterial (Blenk et al. 1986) infections as well as subtyping of the infection agents (Yehle et al. 1987) should be possible not why in sera (Seibl et al. 1989) or tissue streaks, but also with cultured cells (Heiles et al. 1988). Moreover, the new system may be used as a nonradioactive tool in the elucidation of structure and function of oncogenes (Reddy et al. 1982; Tabin et al. 1982; Bishop 1983; Taporowsky et al. 1983; Bishop 1985; Barbacid 1986; Bishop 1986; Der et al. 1986; Sukumar and Barbacid 1986; Bishop 1987) and its correlation to tumor formation (Slamon et al. 1984; Ascione et al. 1986; Klein and Klein 1986; Watson 1986). Besides cytogenetic studies (Sandberg et al. 1988) using pulse-field gel electrophoresis and nonradioactively-labeled DNA probes, DNA rearrangements, including oncogene loci which result in different tumors, may be studied on a molecular level

(Hagemeijer et al. 1985; Rubin et al. 1986; Baer et al. 1987; Hermans et al. 1987; Mengle-Graw and Rabbitts 1987; Rubin et al. 1987).

The new system should also be helpful in mutagenesis research for the detection of tumor-correlated, naturally occurring, or experimentally induced point mutations, by application of digoxigenin-labeled oligonucleotide probes specific for the point mutation of interest. It may be also applied for the localization of genetic defects on the human genome (Stanbury et al. 1983; Davies 1986; LeGrys et al. 1987; Watkins 1988). Examples of important genetic defects under investigation are genetic mapping of X-linked mental retardation (Patterson et al. 1987; Arveiler et al. 1988), carrier detection and prenatal diagnosid of Duchenne muscular dystrophy (Den Dunnen et al. 1987; Kenwrick et al. 1987), diagnosis of hemoglobinopathies and hemophilia (Amselem et al. 1988), the molecular genetics of cystic fibrosis (Estvill et al. 1987) as well as Alzheimer's disease (Kang et al. 1987; St. George-Hyslop et al. 1987) and Huntington's disease (Gilliam et al. 1987). With appropriate DNA markers, the analysis of the evolution of genomes and genetic variation should also be facilitated (Bernardi and Bernardi 1986; Cavalli-Sforza et al. 1986).

Finally, the new digoxigenin system might be used for analyzing vaccines like hepatitis-B antigens produced by genetic engeneering techniques (Murray 1988), or proteins of human origin used in therapy such as tissue-type plasminogen activator (Pennica et al. 1983), erythropoetin (Lin et al. 1985) or clotting factor XIII (Wood et al. 1984) for the absence of nucleic acid contaminations, without having to deal with radioactivity (Slater 1988).

Acknowledgements. The author wishes to thank H.-J. Höltke and R. Seibl for support and critical reading the manuscript, P. Hirth and G. Schumacher for application of the novel system and R. Falter for typing the manuscript.

References

Al-Hakim AH, Hull R (1986) Studies towards the development of chemically synthesized non-radioactive biotinylated nucleic acid hybridization probes. Nucleic Acids Res 14: 9965–9976

Ali S, Müller CR, Epplen JT (1986) DNA finger printing by oligonucleotide probes specific for simple repeats. Hum Genet 74: 239–243

Ambros PF, Matzke MA, Matzke AJM (1986) Detection of a 17 kb unique sequence (T-DNA) in plant chromosomes by in situ hybridization. Chromosoma (Berl) 94: 11–18

Amselem S, Nunes V, Viaud M, Estivill X, Wong C, D'Auriol L, Vidaud D, Galibert F, Baiget M, Goosens M (1988) Determination of the spectrum of β-thalassemia genes in Spain by use of dot-blot analysis of amplified β-globin DNA. Am J Hum Genet 43: 95–100

Anderson G, Deinard AS (1974) Nitroblue tetrazolium (NBT) test. Am J Med Technol 40: 345–353

Anderson MLM, Young BD (1985) Quantitative filter hybridization. In: Hames BD, Higgins SJ (eds) Nucleic acid hybridization: a practical approach. IRL Press, Oxford Washington DC, pp 73–111

Arveiler B, Oberle I, Vincent A, Hofker MH, Pearson PL, Mandel JL (1988) Genetic mapping of the Xq27-q28 region: New RFLP markers useful for diagnostic applications in fragile-X and haemophilia-B families. Am J Hum Genet 42: 380–389

Ascione R, Sacchi N, Watson DK, Fisher RJ, Fujiwara S, Seth A, Papas TS (1986) Oncogenes: molecular probes for clinical application in malignant disease. Gene Anal Techn 3: 25–39

Ausubel FM, Brent R, Kingston RE, Moore DD, Smith JA, Seidman JG, Struhl K (1987) Current protocols in molecular biology. Greene Publ Assoc Wiley-Intersci, NY

Baer R, Forster A, Rabbitts TH (1987) The mechanism of chromosome 14 inversion in human T cell lymphoma. Cell 50: 97–105

Baker CC, Phelps WC, Lindgren V, Braun MJ, Gonda MA, Howley PM (1987) Structural and transcriptional analysis of human papillomavinus type 16 sequences in cerivical carcinoma cell lines. J Virol 61: 962–971

Baker PS (1969) Nuclear data. In: Wang Y (ed) Handbook of radioactive nuclides. CRC Crit Rev, Cleveland Ohio, pp 3–63

Barbacid M (1986) Human oncogenes. Rev Clin Esp 179: 412–427

Beck S (1987) Colorimetric-detected DNA sequencing. Anal Biochem 164: 514–520

Benham FJ, Spurr N, Povey S, Brinton BT, Goodfellow PN, Solomon E, Harris TJR (1984) Assignment of tissue-type plasminogen activator to chromosome 8 in man and identification of a common restriction length polymorphism within the gene. Mol Biol Med 2: 251–259

Benton WD, Davis RW (1978) Screening λgt recombinant clones by hybridization to single plaques in situ. Science 196: 180–182

Bernardi G, Bernardi G (1986) The human genome and its evolutionary context. Cold Spring Harbor symposia on quantitative biology: molecular biology of *Homo sapiens,* vol 51. Cold Spring Harbor Lab, Cold Spring Harbor NY, pp 479–487

Beyer H (1973) Lehrbuch der organischen Chemie, 17. Auflage. Hirzel Stuttgart, pp 610–612

Bilai VI, Shcherbina SM (1977) Dynamics of vitamin content in different *Fusarum* strains. Mikrobiol Zh 39: 597–600

Bishop JM (1983) Cellular oncogenes and retroviruses. Annu Rev Biochem 52: 301–354

Bishop JM (1985) Viral oncogenes. Cell 42: 23–38

Bishop JM (1986) Oncogenes and proto-oncogenes. J Cell Physiol 4: 1–5

Bishop JM (1987) The pathobiology of proto-oncogenes. Proc Int Symp Princess Takamatsu Cancer Res Fund 1986: 3–9

Blenk H, Kuwert T, Zöller L (1986) Diagnosis of *Chlamydial* infections in sexually transmitted diseases. Labmedica 3: 23–28

Böck A, Wirth R, Schmid G, Schumacher G, Lang G, Buckel P (1983a) The two subunits of penicillin acylase are processed from a common precursor. FEMS Microbiol Lett 20: 135–139

Böck A, Wirth R, Schmid G, Schumacher G, Lang G, Buckel P (1983b) The penicillin acylase from *Escherichia coli* ATCC 11105 contains two dissimilar subunits. FEMS Microbiol Lett 20: 141–144

Brigati DJ, Myerson D, Leary JJ, Spalholz B, Travis SZ, Fong CKY, Hsiung GD, Ward DC (1983) Detection of viral genomes in cultured cells and paraffin-embedded tissue sections using biotin-labeled hybridization probes. Virology 126: 32–50

Brown DM, Frampton J, Goelet P, Karn J (1982) Sensitive detection of RNA using strand-specific M13 probes. Gene (Amst) 20: 139–144

Cantor CR, Schimmel PR (1980) Volumes and mass densities of proteins. In: The conformation of biological macromolecules, part 1. Freeman San Francisco, pp 115–117

Cavalli-Sforza LL, Kidd JR, Kidd KK, Bucci C, Bowcock AM, Hewlett BS, Friedlaenders JS (1986) DNA markers and genetic variation in the human species. Cold Spring Harbor symposia on quantitative biology: Molecular biology of *Homo sapiens,* vol 51. Cold Spring Harbor Lab, Cold Spring Harbor NY, pp 411–417

Chaiet L, Wolf FJ (1964) The properties of streptavidin, a biotin-binding protein produced by *Strepomycetes.* Arch Biochem Biophys 106: 1–5

Chien A, Edgar DB, Trela JM (1976) Deoxyribonucleic acid polymerase from the extreme thermophile *Thermus aquaticus.* J Bacteriol 127: 1550–1557

Church GM, Gilbert W (1984) Genomic sequencing. Proc Natl Acad Sci USA 81: 1991–1995

Cook AF, Vuocolo E, Brakel CL (1988) Synthesis and hybridization of a series of biotinylated oligonucleotides. Nucleic Acids Res 16: 4077–4095

Craviato A, Trujillo F, Beltran P, Hill WE (1988) DNA hybridization with oligodeoxyribonucleotide probes for identifying enterotoxin-producing *Escherichia coli*. Mol Cell Probes 2: 125–130

Dahlen P, Syvänen AC, Hurskainen P, Kwiatkowski M, Sund C, Ylikoski J, Söderlund H, Lövgren T (1987) Sensitive detection of genes by sandwich hybridization and time-resolved fluorometry. Mol Cell Probes 1: 159–168

Davie EW, Ichinose A, Leytus SP (1986) Structural features of the proteins participating in blood coagulation and fibrinolysis. Cold Spring Harbor symposia on quantitative biology: Molecular biology of *Homo sapiens*, vol 51. Cold Spring Harbor Lab, Cold Spring Harbor NY, pp 509–514

Davies KE (1986) Human genetic diseases. A practical approach. IRL Press, Oxford Washington DC

Delius H, Gough NM, Cameron CH, Murray K (1983) Structure of the hepatitis B genome. J Virol 47: 337–343

DeLuca M, McElroy WD (1978) Purification and properties of firefly luciferase. In: Colowick SP, Kaplan NO (eds) Methods enzymol, vol 57. Academic Press, Lond NY, pp 3–15

Den Dunnen JT, Bakker E, Klein Breteler EG, Pearson PL, Van Ommen GJB (1987) Direct detection of more than 50% of the Duchenne muscular dystrophy mutations by field inversion gels. Nature (Lond) 329: 640–642

Deng G, Wu R (1983) Terminale transferase: Use in the tailing of DNA and for in vitro mutagenesis. In: Wu R, Grossmann L, Moldave K (eds) Methods enzymol, vol 100. Academic Press, Lond NY, pp 96–116

Denhardt D (1966) A membrane-filter technique for the detection of complementary DNA. Biochem Biophys Res Comm 23: 641–646

Der CJ, Finkel T, Cooper GM (1986) Biological and biochemical properties of human *ras*^{H} genes mutated at codon 61. Cell 44: 167–176

Donis-Keller H, Green P, Helms C, Cartinhour S, Weiffenbach B, Stephens K, Keith TP, Bowden DW, Smith DR, Lander ES, Botstein D, Akots G, Rediker KS, Gravius T, Brown VA, Rising MB, Parker C, Powers JA, Watt DE, Kauffman ER, Bricker A, Phipps P, Muller-Kahle H, Fulton TR, Ng S, Schumm JW, Braman JC, Knowiton RG, Barker DF, Crooks SM, Lincoln SE, Daly MJ, Abrahamson J (1987) A genetic linkage map of the human genome. Cell 51: 319–337

Donovan RM, Bush CE, Peterson WR, Parker LH, Cohen SH, Jordan GW, Vanden Brink KM, Goldstein E (1987) Comparison of non-radioactive DNA hybridization probes to detect human immunodeficiency virus nucleic acid. Mol Cell Probes 1: 359–366

Dykes DD (1988) The use of biotinylated DNA probes in parentage testing: non-isotopic labeling and non-toxic extraction. Electrophoresis 9: 359–368

Eggers HJ (1977) Selective inhibition of uncoating of echovirus 12 by rhodamine. Virology 78: 241–252

Epanchinov AV (1976) Accumulation of vitamins in the corn radical zone after soil fertilization. Prikl Biochim Mikrobiol 12: 96–102

Estvill X, Scambler PJ, Wainwright BJ, Hawley K, Frederick P, Schwarz M, Baiget M, Kere J, Williamson R, Farrall M (1987) Patterns of polymorphism and linkage disequilibrium for cystic fibrosis. Genomics 1: 257–263

Feinberg AP, Vogelstein B (1983) A technique for radiolabeling DNA restriction endonuclease fragments to high specific activity. Anal Biochem 132: 6–13

Feinberg AP, Vogelstein B (1984) Addendum: a technique for radiolabeling DNA restriction endonuclease fragments to high specific activity. Anal Biochem 137: 266–267

Fisher R, Waller EK, Grossi G, Thompson D, Tizard R, Schleunig WD (1985) Isolation and characterization of the human tissue-type plasminogen activator structual gene including its 5‘ flanking region. J Biol Chem 260: 11223–11230

Forster AC, McInnes JL, Skingle DC, Symons RH (1985) Non-radioactive hybridization probes prepared by the chemical labelling of DNA and RNA with a novel reagent, photobiotin. Nucleic Acids Res 13: 745–761

Franzi C, Vidal J (1988) Coupling redox and enzymic reactions improves the sensitivity of the ELISA-spot assay. J Immunol Methods 107: 239–244

Gall JG, Pardue ML (1969) Formation and detection of RNA-DNA hybrid molecules in cytological preparations. Proc Natl Acad Sci USA 63: 378–383

Gall JG, Pardue ML (1971) Nucleic acid hybridizations in cytological preparations. In: Colowick SP,Kaplan NO (eds) Methods enzymol, vol 21. Academic Press, Lond NY, pp 470–480

Garbutt GJ, Wilson JT, Schuster GS, Leary JJ, Ward DC (1985) Use of biotinylated probes for detecting sickle cell anemia. Clin Chem 31: 1203–1206

Gebeyehu G, Rao PY, SooChan P, Simms DA, Klevan L (1987) Novel biotinylated nucleotide-analogues for labeling and colorimetric detection of DNA. Nucleic Acids Res 15: 4513–4534

Georges M, Lequarre AS, Castelli M, Hanset R, Vassart G (1988) DNA fingerprinting in domestic animals using four different minisatellite probes. Cytogenet Cell Genet 47: 127–131

Gill P, Jeffreys AJ, Werrett DJ (1985) Forensic application of DNA fingerprints. Nature (Lond) 318: 577–579

Gillespie D, Spiegelman S (1965) A quantitative assay for DNA-RNA hybrids with DNA immobilized on a membrane. J Mol Biol 12: 829–842

Gilliam IC, Tener GM (1986) N^4-(6-Aminohexyl)cytidine and -deoxycytidine nucleotides can be used to label DNA. Anal Biochem 157: 199–207

Gilliam TC, Tanzi RE, Haines JL, Bonner JI, Faryniarz AG, Hobbs WJ, MacDonald ME, Cheng SV, Folstein SE, Conneally PM, Wexler NS, Gusella JF (1987) Localization of the Huntington's disease gene to a small segment of chromosome 4 flanked by *D4S10* and the telomere. Cell 50: 565–571

Glover DM (1985a) DNA cloning. A practical approach, vol 1. IRL Press, Oxford Washington DC

Glover DM (1985b) DNA cloning. A practical approach, vol 2. IRL Press, Oxford Washington DC

Glover DM (1987) DNA cloning. A practical approach, vol 3. IRL Press, Oxford Washington DC

Green NM (1975) Avidin. Adv Protein Chem 29: 85–133

Gregersen N, Koch J, Kolvraa S, Petersen KB, Bolund L (1987) Improved methods for the detection of unique sequences in Southern blots of mammalian DNA by non-radioactive biotinylated DNA hybridization probes. Clin Chim Acta 169: 267–280

Gross GI (1987) Viroids. Infectious nucleic acids: the pathogens of plant diseases – a mini-review. Mol Biol 21: 1480–1485

Gross HJ, Krupp G, Domdey H, Raba M, Jank P, Lossow C, Alberty H, Ramm K, Sanger HL (1982) Nucleotide sequence and secondary structure of citrus exocortis and chrysanthemum stunt viroid. Eur J Biochem 121: 249–257

Grunstein M, Hogness DS (1975) Colony hybridization. Method for the isolation of cloned DNA's that contain a specific gene. Proc Natl Acad Sci USA 72: 3961–3965

Gundermann KD, Horstmann W, Bergmann G (1965) Konstitution und Chemilumineszenz II. Synthese und Chemilumineszenz-Verhalten von 7-Dialkylamino-Naphtalin-Dicarbonsäure-(1.2)-Hydraziden. Liebigs Ann Chem 684: 127–141

Guo LH,Yang RCA, Wu R (1983) An improved strategy for rapid direct sequencing of both strands of long DNA molecules cloned in a plasmid. Nucleic Acids Res 11: 5521–5540

Gupta JW, Gupta PK, Rosenshein N, Shah KV (1978) Detection of human papilloma virus in cervical smears. Arch Cytol 31: 387–396

Haas MJ and Fleming DJ (1988) A simplified lysis method allowing the use of biotinylated probes in colony hybridization. Anal Biochem 168: 239–246

Hagemeijer A, DeKlein A, Gödde-Salz E, Turc-Carel C, Smit EME, Van Agthoven AJ, Grosveld GC (1985) Translocation of c-*abl* to "masked" Ph in chronic myeloid leukemia. Cancer Gen Cytogen 18: 95–104

Harris TJR, Patel T, Marston FAO, Little S, Emtage JS, Opdenakker G, Volckaert G, Rombauts W, Billian A, De Somer P (1986) Cloning of cDNA coding for human tissue-type plasminogen activator and its expression in *Escherichia coli*. Mol Biol Med 3: 279–292

Hauber R, Geiger R (1987) A new, very sensitive, bioluminescence-enhanced detection system for protein blotting. I. Ultrasensitive detection systems for protein blotting and DNA hybridization. J Clin Chem Clin Biochem 25: 511–514

Hegnauer R (1971) Pflanzenstoffe und Pflanzensystematik. Naturwissenschaften 58: 585–598

Heiles BJ, Genersch E, Kessler C, Neumann R, Eggers HJ (1988) In situ hybridization with digoxigenin-labeled DNA of human papillomaviruses (HPV 16/18) on HeLa and SiHa cells. BioTechniques 6: 978–981

Hermans A, Heisterkamp N, Von Lindern M, Van Baal S, Meijer D, Van Der Plas D, Wiedemann LM, Groffen J, Bootsma D, Grosveld G (1987) Unique fusion of *bcr* and c-*abl* genes in Philadelphia chromosome positive acute lymphoblastic leukemia. Cell 51: 33–40

Herzberg M (1984) Molecular genetic probe, assay technique, and a kit using this molecular genetic probe. Eur Pat Appl 0 128 018 A2, pp 1–42

Höfler H (1987) What's new in »In situ hybridization«. Pathol Res Pract 182: 421–430

Holck AR, Meek CL (1987) Dose-mortality responses of crawfish and mosquitoes to selected pesticides. J Am Mosq Control Assoc 3: 407–411

Höltke HJ, Seibl R, Burg J, Mühlegger K, Kessler C (1989) Non-radioactive DNA labeling and detection of nucleic acids: II: Optimizations of the digoxigenin system. Nucleic Acids Res (submitted)

Hood LE, Hunkapiller MW, Smith LM (1987) Automated DNA sequencing and analysis of the human genome. Genomics 1: 201–212

Hopman AHN, Wiegant J, Van Duijn P (1987) Mercurated nucleic acid probes, a new principle for non-radioactive in situ hybridization. Exp Cell Res 169: 357–368

Horwitz JP, Chua J, Noel M, Donatti JT, Freisler J (1966) Substrates for cytochemical demonstration of enzyme activity. II.Some dihalo-3-indolyl phosphates and sulfates. J Med Chem 9: 447

Hunter MM, Margolies MN, Ju A, Haber E (1982) High-affinity monoclonal antibodies to the cardiac glycoside, digoxin. J Immunol 129: 1165–1172

Hyman HC, Yogev D, Razin S (1987) DNA probes for detection and identification of *Mycoplasma pneumoniae* and *Mycoplasma genitalium*. J Clin Microbiol 25: 726–728

Inoue H, Gushi K, Matsuura S, Sakata Y (1987) A sensitive colorimetric detection of virus DNA and oncogene. Biochem Biophys Res Comm 143: 323–328

Inove S, Hashida S, Tanaka K, Imagawa M, Ishikawa E (1985) Preparation of monomeric affinity-purified Fab'–β–D–galactosidase conjugate for immunoenzymometric assay. Anal Lett 18: 1331–1344

Ishikawa E, Imagawa M, Hashida S, Yoshitake S, Hamaguchi Y, Keno T (1983) Enzyme–labeling of antibodies and their fragments for enzyme immunoassay and histochemical staining. J Immunoassay 4: 209–327

Jablonski E, Moomaw EW, Tullis RH, Ruth JL (1986) Preparation of oligodeoxynucleotide-alkaline phosphatase conjugates and their use as hybridization probes. Nucleic Acids Res 14: 6115–6128

Jeffreys AJ, Brookfield JFY, Semeonoff R (1985a) Positive identification of an immigration test-case using human DNA fingerprints. Nature (Lond) 317: 818–819

Jeffreys AJ, Wilson SL, Thein SL (1985b) Hypervariable »minisatellite« regions in human DNA. Nature (Lond) 314: 67–73

Kafatos FC, Jones CW, Efstratiadis A (1979) Determination of nucleic acid sequence homologies and relative concentrations by a dot hybridization procedure. Nucleic Acids Res 7: 1541–1552

Kahl G (1988) Architecture of eukaryotic genes. VCH Verlagsgesellschaft Weinheim

Kaledin AS, Slyusarenko AG, Gorodetskii SI (1980) Isolation and properties of DNA polymerase from extremal thermophylic bacteria *Thermus aquaticus* YT-1. Biokhymiya 45: 644–651

Kamdar H, Jayaraman K (1983) Spontaneous loss of a high-molecular-weight plasmid and the biocide of *Bacillus thuringiensis* var. israelensis. Biochem Biophys Res Comm 110: 477–482

Kandolf R, Ameis D, Kirschner P, Canu A, Hofschneider PH (1987) In situ detection of enteroviral genomes in myocardial cells by nucleic acid hybridization: an approach to the diagnosis of viral heart disease. Proc Natl Acad Sci USA 84: 6272–6276

Kang J, Lemaire HG, Unterbeck A, Salbaum JM, Masters CL, Grzeschik KH, Multhaup G, Beyreuther K, Müller-Hill B (1987) The precursor of Alzheimer's disease amyloid A4 protein resembles a cell-surface receptor. Nature (Lond) 325: 733–736

Keese P, Symons RH (1985) Domains in viroids: evidence of intermolecular RNA rearrangements and their contribution to viroid evolution. Proc Natl Acad Sci USA 82: 4582–4586

Keller GH, Cumming CU, Huang DP, Manak MM, Ting R (1988) A chemical method for introducing haptens onto DNA probes. Anal Biochem 170: 441–450

Kenwrick S, Patterson M, Speer A, Fischbeck K, Davies K (1987) Molecular analysis of the Duchenne muscular dystrophy region using pulsed field gel electrophoresis. Cell 48: 351–357

Kessler C, Höltke HJ, Seibl R, Burg J, Mühlegger K, (1989) Non-radioactive DNA labeling and detection of nucleic acids: I. A novel highly sensitive system based on digoxigenin:anti-digoxigenin ELISA principle. Nucleic Acids Res (submitted)

Khandjian EW (1987) Optimized hybridization of DNA blotted and fixed to nitrocellulose and nylon membranes. BioTechnology 5: 165–167

Klein G, Klein E (1986) Conditioned tumorigenicity of activated oncogenes. Cancer Res 46: 3211–3224

Kristensen T, Voss H, Schwager C, Sproat B, Ansorge W (1988) T7 DNA polymerase in automated dideoxy sequencing. Nucleic Acids Res 16: 3487–3496

Lacey E, Grant WN (1987) Photobiotin as a sensitive probe for protein labeling. Anal Biochem 163: 151–158

Langer PR, Waldrop AA, Ward DC (1981) Enzymatic synthesis of biotin-labeled polynucleotides: novel nucleic acid affinity probes. Proc Natl Acad Sci USA 78: 6633–6637

Langer-Safer PR, Levine M, Ward DC (1982) Immunological method for mapping genes on *Drosophila* polytene chromosomes. Proc Natl Acad Sci USA 79: 4381–4385

Lardy HA, Peanasky R (1953) Metabolic functions of biotin. Physiol Rev 33: 560–565

Leary JJ, Brigati DJ, Ward DC, (1983) A sensitive colorimetric method for visualizing biotin-labeled probes hybridized to DNA or RNA on nitrocellulose filters. Chromosomes and Cancer, chapt 15. Academic Press, Lond NY, pp 273–290

LeGrys VA, Leinbach SS, Silverman LM (1987) Clinical applications of DNA probes in the diagnosis of genetic diseases. CRC Crit Rev Clin Lab Sci 25: 255–274

Lin FK, Suggs S, Lin CH, Browne JK, Smalling R, Egrie JC, Chen KK, Fox GM, Martin F, Stabinsky Z, Badrawi SM, Lai PH, Goldwasser E (1985) Cloning and expression of the human erythropoietin gene. Proc Natl Acad Sci USA 82: 7580–7584

Maclean N,Gregory SP, Flavell RA (1983) Eukaryotic genes. Their structure, activity and regulation. Butterworth, Lond

Magnius LO, Espmark JA (1972) New specificities in Australian antigen positive sera distinct from the Le Bouvier determinants. J Immunol 109: 1017

Maitland NJ, Cox MF, Lynas C, Prime S, Crane I, Scully C (1987) Nucleic acid probes in the study of latent viral disease. J Oral Pathol 16: 199–211

Maniatis T, Fritsch EF, Sambrook J (1982) Molecular cloning. A laboratory manual. Cold Spring Harbor Lab, Cold Spring Harbor NY

Manuelidis L, Langer-Safer PR, Ward DC (1982) High-resolution mapping of satellite DNA using biotin-labeled DNA probes. J Cell Biol 95: 619–625

Martin R, Hoover C, Grimme S, Grogan C, Höltke HJ, Kessler C (1988) Applications of non-radioactive digoxigenin labeling and detection system. BioTechniques (submitted)

Mather MW (1988) Base composition-independent hybridization in dried agarose gels: screening and recovery for cloning of genomic DNA fragments. BioTechniques 6: 444–447

Matthews JA, Batki A, Hynds C, Kricka LJ (1985) Enhanced chemiluminescent method for the detection of DNA dot-hybridization assays. Anal Biochem 151: 205–209

Maxam AM, Gilbert W (1977) A new method for sequencing DNA. Proc Natl Acad Sci USA 74: 560–564

Maxam AM, Gilbert W (1980) Sequencing end-labeled DNA with base-specific chemical cleavages. In: Grossman L, Moldave K (eds) Methods enzyme, vol 65. Academic Press Lond NY, pp 499–560

McGadey J (1970) A tetrazolium method for non-specific alkaline phosphatase. Histochemie 23: 180–184

Meinkoth J, Wahl G (1984) Hybridization of nucleic acids immobilized on sold supports. Anal Biochem 138: 267–284

Mengle-Graw L, Rabbitts TH (1987) A human chromosome 8 region with abnormalities in B cell, HTLV-I$^+$ T cell and c-*myc* amplified tumors. EMBO J 6: 1959–1965

Mincheva A, Gissmann L, Zur Hausen H (1987) Chromosmal integration sites of human papillomavirus DNA in three cervical cancer cell lines mapped by in situ hybridization. Med Microbiol Immunol 176: 245–256

Miyakawa Y, Mayumi M (1981) HB_eAg-Anti-HB_e system in hepatitis B virus infection. In: Szmuness W, Alter HJ, Maynard JE (eds) Viral hepatitis. The Franklin Inst Press, Philadelphia, pp 183–194

Moench TR (1987) In situ hybridization. Mol Cell Probes 1: 195–205

Mühlegger K, Huber E, von der Eltz H, Kessler C (1989) Non-radioactive DNA labeling and detection of nucleic acids/ IV. Chemical synthesis of the nucleotide compounds of the digoxigenin system and photodigoxigenin. Nucleic Acid Res (submitted)

Mullis KB, Faloona FA (1988) Specific synthesis of DNA in vitro via a polymerase-catalyzed chain reaction. In: Colowick SP, Kaplan NO (eds) Methods enzymol, vol 155. Academic Press, Lond NY, pp 335–350

Mullis KB, Faloona FA, Scharf S, Saiki R, Horn G, Erlich H (1986) Specific enzymatic amplification of DNA in vitro with the polymerase chain reaction. Cold Spring Harbor symp quant biol: Molecular biology of *Homo sapiens*, vol 51. Cold Spring Harbor Lab, Cold Spring Harbor NY, pp 263–273

Murray K (1988) Application of recombinant DNA techniques in the development of viral vaccines. Vaccine 6: 164–174

Nakamura Y, Leppert M, O'Connell P, Wolff R, Holm T, Culver M, Martin C, Fujimoto E, Hoff M, Kulmin E, White R (1987) Variable number of tandem repeat (VNTR) markers for human gene mapping. Science 235: 1616–1622

Neumüller OA (1981) Römpps Chemie-Lexikon: Digitalin, 8. Aufl, Bd 2. Franckh'sche Verlagshandlung, Stuttgart, pp 956–958

Ny T, Elgh F, Lund B (1984) The structure of the human tissue-type plasminogen activator gene: correlation of intron and exon structures to functional and structural domains. Proc Natl Acad Sci USA 81: 5355–5359

Obe G,Basler A (1987) Cytogenetics (Basel). Springer V, Berlin Heidelberg New York Tokyo

Opdenakker G, Billau A, Volckaert G, De Somer P (1985) Determination of tissue-type plasminogen-activator mRNA in human and non-human cell lines by dot-blot hybridization. Biochem J 231: 309–313

Oprandy JJ, Thornton SA, Gardiner CH, Burr D, Batchelor R, Bourgeois AL (1988) Alkaline phosphatase-conjugated oligonucleotide probes for enterotoxigenic *Escherichia coli* in travelers to South America and West Africa. J Clin Microbiol 26: 92–95

Oste C (1988) Polymerase chain reaction. BioTechniques 6: 162–167

Ou CY, Kwok S, Mitchell SW, Mack DH, Sninsky JJ, Krebs JW, Feorino P, Warfield D, Schochetman G (1988) DNA amplification for direct detection of HIV–1 in DNA of peripheral blood mononuclear cells. Science 239: 295–297

Pataki S, Meyer K, Reichstein T (1953) Die Konfiguration des Digoxigenins (Teilsynthese des 3β-, 12α- und des 3β,12α-Dioxyätiansäure-methylesters). Glykoside and Aglycone, 116. Mitteilung. Helv Chim Acta 36: 1295–1308

Patterson M, Kenwrick S, Thibodeau S, Faulk K, Mattei MG, Davies KE (1987) Mapping of DNA markers close to the fragile site on the human X chromosome at Xq27.3. Nucleic Acids Res 15: 2639–2651

Pennica D, Holmes WE, Kohr WJ, Harkins RN, Vehar GA, Ward CA, Bennet WF, Yelverton E, Seeburg PH, Heynecker HL, Goeddel DV (1983) Cloning and expression of human tissue-type plasminogen activator cDNA in *E.coli*. Nature (Lond) 301: 214–221

Pereira HG (1986) Non-radioactive nucleic acid probes for the diagnosis of virus infections. BioEssays 4: 110–113

Pezzela M, Pezella F, Galli C, Macchi B, Verani P, Sorice F, Baroni CD (1987) In situ hybridization of human immunodeficiency virus (HTLV-III) in cystostat sections of lymph nodes of lymphadenopathy syndrome patients. J Med Virol 22: 135–142

Pitcher DG, Owen RJ, Dyal P, Beck A (1987) Synthesis of a biotinylated DNA probe to detect ribosomal RNA cistrons in *Providencia stuartii*. FEMS Microbiol Lett 48: 283–287

Popescu NC, DiPaolo JA, Amsbaugh SC (1987) Integration sites of human papillomavirus 18 DNA sequences on HeLa cell chromosomes. Cytogenet Cell Genet 44: 58–62

Porstmann T, Terynck T, Avrameas S (1985) Quantitation of 5-bromo-2-deoxyuridine incorporation into DNA: an enzyme immunoassay for the assessment of the lymphoid cell proliferative response. J Immunol Meth 82: 169–179

Prince AM (1968) An antigen detected in the blood during the incubation period of serum hepatitis human virus. Proc Natl Acad Sci USA 60: 814–821

Puget K, Michelson AM (1976) Microestimation of glucose and glucose oxidase. Biochimie 58: 757–758

Rashtchian A, Eldredge J, Ottaviani M, Abbott M, Mock G, Lovern D, Klinger J, Parsons G (1987) Immunological capture of nucleic acid hybrids and application to nonradioactive DNA probe assay. Clin Chem 33: 1526–1530

Reddy EP, Reynolds RK, Santos E, Barbacid M (1982) A point mutation is responsible for the aquisition of transforming properties by the T24 human bladder carcinoma oncogene. Nature (Lond) 300: 149–152

Redeker AG (1975) Viral hepatitis clinical aspects. Am J Med Sci 270: 9–16

Reichstein T (1962) Besonderheiten der Zucker von herzaktiven Glykosiden. Angew Chem 74: 887–918

Reichstein T, Weiss E (1962) The sugars of the cardiac glycosides. Adv Carbohydr Chem 17: 65–120

Reisfeld A, Rothenberg JM, Bayer EA, Wilchek M (1987) Non-radioactive hybridization probes prepared by the reaction of biotin hydrazide with DNA. Biochem Biophys Res Comm 142: 519–526

Renz M, Kurz C (1984) A colorimetric method for DNA hybridization. Nucleic Acids Res 12: 3435–3444

Richman DD, McCutchan JA, Spector SA (1987) Detecting human immunodeficiency virus RNA in peripheral blood mononuclear cells by nucleic acid hybridization. J Infect Dis 156: 823–827

Riesner D, Gross HJ (1985) Viroids. In: Richardson CC (ed) Ann rev biochem, vol 54. Annu Rev Inc, Palo Alto Cal, pp 531–564

Rigby PWJ, Dieckmann M, Rhodes C, Berg P (1977) Labeling deoxyribonucleic acid to high specific activity in vitro by nick translation with DNA polymerase I. J Mol Biol 113: 237–251

Riley LK, Marshall ME, Coleman MS (1986) A method for biotinylating oligonucleotide probes for use in molecular hybridizations. DNA 5: 333–337

Romanchuk KG (1982) Fluorescein. Physicochemical factors affecting its fluorescence. Surv Ophthalmol 26: 269–283

Rubin CM, LeBeau MM, Smith SD, Rowley JD, Westbrook CA (1986) Philadelphia chromosome-positive (Ph^1+) acute lymphoblastic leukemia (ALL): DNA rearrangements on chromosome 9 and 22 identified using pulse field gel electrophoresis (PFGE). Blood 68: 264a

Rubin CM, Westbrook CA, Rowley JD (1987) Mapping of the c-*abl* locus using pulse field electrophoresis: application to the analysis of chromosome translocations. Am J Hum Genet 41: A215

Ryskov AP, Jincharadze AG, Prosnyak MI, Ivanov PL, Limborska SA (1988) M13 phage DNA as a universal marker for DNA fingerprinting of animals, plants and microorganisms. FEBS Lett 233: 388–392

Saiki RK, Arnheim N, Erlich HA (1985a) A novel method for the detection of polymorphic restriction sites by cleavage of oligonucleotide probes: application to sickle-cell anemia. BioTechnology 3: 1008–1012

Saiki RK, Scharf S, Faloona F, Mullis KB, Horn GT, Erlich HA, Arnheim N (1985b) Enzymatic amplification of β-globin sequences and restriction site analysis for diagnosis of sickle cell anemia. Science 230: 1350–1354

Saiki RK, Bugawan TL, Horn GT, Mullis KB, Erlich HA (1986) Analysis of enzymatically amplified β-globin and HLA-DQα DNA with allele-specific oligonucleotide probes. Nature (Lond) 324: 163–166

Saman E (1986) A simple and sensitive method for detection of nucleic acids fixed on nylon-based filters. Gene Anal Techn 3: 1–5

Sandberg AA (1980) The chromosomes in human cancer and leukemia. Elsevier North Holland Biomed Press, Amst NY

Sandberg AA, Turc-Carel C, Gemmill RM (1988) Chromosomes in solid tumors and beyond. Cancer Res 48: 1049–1059

Sanger F, Nicklen S, Coulson AR (1977) DNA sequencing with chain-terminating inhibitors. Proc Natl Acad Sci USA 74: 5463–5467

Sanger F, Coulson AR, Barrell BG, Smith AJH, Roe BA (1980) Cloning in single-stranded bacteriophage as an aid to rapid DNA sequencing. J Mol Biol 143: 161–178

Sarkar G, Sommer SS (1988) RNA amplification with transcript sequencing (RAWTS). Nucleic Acids Res 16: 5197

Schäfer R, Zischler H, Epplen JT (1988a) $(CAC)_5$, a very informative oligonucleotide probe for DNA fingerprinting. Nucleic Acids Res 16: 5196

Schäfer R, Zischler H, Epplen JT (1988b) DNA fingerprinting using non-radioactive oligonucleotide probes specific for simple repeats. Nucleic Acids Res 16: 9344

Schäfer R, Zischler H, Birser U, Becker A, Epplen JT (1988c) Optimized oligonucleotide probes for DNA fingerprinting. Electrophoresis 9: 369–374

Schumacher G, Sizman D, Haug H, Buckel P, Böck A (1986) Penicillin acylase from *E. coli*. Unique gene-protein relation. Nucleic Acids Res 14: 5713–5727

Schwarz E, Freese UK, Gissmann L, Mayer W, Roggenbuck B, Zur Hausen H (1985) Structure and transcription of human papillomavirus sequences in cervical carcinoma cells. Nature (Lond) 314: 111–114

Seibl R, Höltke HJ, Rüger R, Meindl J, Zachau HG, Raßhofer R, Roggendorf M, Wolf H, Arnold N, Wienberg J, Kessler C (1989) Non-radioactive DNA labeling and detection of nucleic acids: III. Applications of the digoxigenin system. Nucleic Acids Res (submitted)

Seriwatana J, Echeverria P, Taylor DN, Sakuldaipeara T, Changchawalit S, Chivoratanond O (1987) Identification of enterotoxigenic *Escherichia coli* with synthetic alkaline phosphatase-conjugated oligonucleotide DNA probes. J Clin Microbiol 25: 1438–1441

Shimizu M, Oshie K, Nakamura K, Takada Y, Oeda K, Ohkawa H (1988) Cloning and expression in *Escherichia coli* of the 135-kDa insecticidal protein gene from *Bacillus thuringiensis* subsp *aizawai* IPL7. Agric Biol Chem 52: 1565–1573

Singer RH, Ward DC (1982) Actin gene expression visualized in chicken muscle tissue culture by using in situ hybridization with a biotinated nucleotide analog. Proc Natl Acad Sci USA 79: 7331–7335

Slamon DJ, DeKernion JB, Verma IM, Cline MJ (1984) Expression of cellular oncogenes in human malignancies. Science 224: 256–262

Slater M (1988) Regulatory affairs issues in biotechnology. Drug Inf J 22: 123–131

Southern EM (1975) Detection of specific sequences among DNA fragments separated by gel electrophoresis. J Mol Biol 98: 503–517

Stanbury JB, Wyngaarden JB, Frederickson DS, Goldstein JL, Brown MS (1983) The metabolic basis of inherited disease, 5th edn. McGraw-Hill NY, pp 39–59

St George-Hyslop PH, Tanzi RE, Polinsky RJ, Haines JL, Nee L, Watkins PC, Myers RH, Feldman RG, Pollen D, Drachman D, Growdon J, Bruni A, Foncin JF, Salmon D, Frommelt P, Amaducci L, Sorbi S, Piacentini S, Stewart GD, Hobbs WJ, Conneally PM, Gusella JF (1987) The genetic defect causing familial Alzheimer's disease maps on chromosome 21. Science 235: 885–890

Stoflet ES, Koeberl DD, Sarkar G, Sommer SS (1988) Genomic amplification with transcript sequencing. Science 239: 491–494

Stute R (1980) The serological diagnosis of hepatitis B - order of importance of hepatitis B markers. Hyg Med 5: 323–325

Su TS, Lin LH, Chou CK, Chang C, Ting LP, Hu CP, Han SH (1986) Hepatitis B virus transcripts in a human hepatoma cell line, HEP 3B. Biochem Biophys Res Comm 138: 131–138

Sukumar S, Barbacid M (1986) The role of *ras* oncogenes in chemically-induced tumors. Pontif Acad Sci Scr Varia 70: 35–53

Swierkosz EM, Scholl DR, Brown JL, Jollick JD, Gleaves CA (1987) Improved DNA hybridization mehtod for detection of acylovir-resistant herpex simplex virus. Antimicrob Agents Chemother 31: 1465–1469

Syvänen AC, Tchen P, Ranki M, Söderlund H (1986) Time-resolved fluorometry: a sensitive method to quantify DNA-hybrids. Nucleic Acids Res 14: 1017–1028

Tabin CJ, Bradley SM, Bargmann CI, Weinberg RA, Papageorge AG, Scolnick EM, Dhar R, Lowy DR, Change EH (1982) Mechanism of activation of a human oncogene. Nature (Lond) 300: 143–149

Taparowsky E, Shimizu K, Goldfarb M, Wigler M (1983) Structure and activation of the human N-*ras* gene. Cell 34: 581–586

Telford JL, Kressmann A, Koski RA, Grosschedl R, Mueller F, Clarkson SG, Birnstiel ML (1979) Delimitation of a promoter for RNA polymerase III by means of a functional test. Proc Natl Acad Sci USA 76: 2590–2594

Theveny B, Revet B (1987) DNA orientation using specific avidin-ferritin biotin end labelling. Nucleic Acids Res 15: 947–958

Ticehurst JR, Feinstone SM, Chestnut T, Tassopoulos NC, Popper H, Purcell RH (1987) Detection of hepatitis A virus by extraction of viral RNA and molecular hybridization. J Clin Microbiol 25: 1822–1829

Tomlinson S, Lyga A, Huguenel E, Dattagupta N (1988) Detection of biotinylated nucleic acid hybrids by antibody-coated gold colloid. Anal Biochem 171: 217–222

Toneguzzo F, Glynn S, Levi E, Mjolsness S, Hayday A (1988) Use of chemically modified T7 DNA polymerase for manual and automated sequencing of supercoiled DNA. BioTechniques 6: 460–469

Traincard F, Terynck T, Danchin A, Avrameas S (1983) Une technique immunoenzymatique pour la mise en evidence de l'hybridation moleculaire entre acides nucleiques. Ann Immunol 134D: 399–405

Tu CPD, Cohen S (1980) 3'-end labeling of DNA with [α-^{32}P]cordycepin-5'-triphosphate. Gene (Amst) 10: 177–183

Urdea MS, Warner BD, Running JA, Stempien M, Clyne J, Horn T (1988) A comparison of non-radioisotopic hybridization assay methods using fluorescent, chemiluminescent and enzyme-labeled synthetic oligodeoxyribonucleotide probes. Nucleic Acids Res 16: 4937–4956

Vaeck M, Botterman J, Reynaerts A, De Block M, Hofte H, Leemans J (1987) Engeneering improved crops for agriculture: Protection from insects and resistance to herbicides. UCLA Symp Mol Cell Biol, New Ser 62: 171–181

Vary CPH, McMahon FJ, Barbone FP, Diamond SE (1986) Non-isotopic detection methods for strand displacement assays of nucleic acids. Clin Chem 32: 1696–1701

Vassart G, Georges M, Monsieur R, Brokas H, Lequarre AS, Christophe D (1987) A sequence in M13 phage detects hypervariable minisatellites in human and animal DNA. Science 235: 683–684

Verschueren H (1985) Interference reflection microscopy in cell biology: Methodology and applications. J Cell Sci 75: 279–301

Viscide RP, Yolken RH (1987) Molecular diagnosis of infectious diseases by nucleic acid hybridization. Mol Cell Probes 1: 3–14

Viscidi RP, Connelly CJ, Yolken RH (1986) Novel chemical method for the preparation of nucleic acids for nonisotopic hybridization. J Clin Microbiol 23: 311–317

Voss EWJr (1984) Immunological properties of fluorescein. In: Voss EWJr (ed) Fluorescein hapten: Immunol probe. CRC Crit Rev, Boca Raton Fla, pp 3–14

Wallace RB, Johnson PF, Tanaka S, Schöld M, Itakura K, Abelson J (1980) Direct deletion of a yeast transfer RNA intervening sequence. Science 209: 1396–1400

Watanabe K, Yano SI, Yamada Y (1982) The selection of cultured plant cell lines producing high levels of biotin. Phytochemistry 21: 513–516

Watkins PC (1988) Restriction fragment length polymorphism (RFLP): applications in human chromosome mapping and genetic disease research. BioTechniques 6: 310–322

Watson JV (1986) Oncogenes, cancer and analytical cytology. Cytometry 7: 400–410

Watson JV (1987) Quantitation of molecular and cellular probes in populations of single cells using fluorescence. Mol Cell Probes 1: 121–136

Weidle UH, Buckel P, Wienberg J (1988) Amplified expression constructs for human tissue-type plasminogen activator in chinese hamster ovary cells: instability in the absence of selection pressure. Gene (Amst) 66: 193–203

Weinberg RA (1985) The action of oncogenes in the cytoplasm and nucleus. Science 230: 770–776

White R, Leppert M, Bishop DT, Barker D, Berkowitz J, Brown C, Callahan P, Holm T, Jerominski L (1985) Construction of linkage maps with DNA markers for human chromosomes. Nature (Lond) 313: 101–105

Wilchek M, Bayer EA (1988) The avidin-biotin complex in bioanalytical applications. Anal Biochem 171: 1–32

Wilson MB, Nakane PK (1978) Recent developments in the periodate mehtod of conjugating horseradish peroxidase to antibodies. In: Knapp W, Holubar K, Wick G (eds) Immunofluorescence and related staining techniques. Elsevier North-Holland Biomed Press, NY Amst, pp 215–223

Wolf H (1987) Molecular biology methods in the diagnosis of viral disease. Experientia (Basel) 43: 1189–1192

Wong C, Dowling CE, Saiki RK, Higuchi RG, Erlich HA, Kazazian HH (1987) Characterization of β-thalassemia mutations using direct genomic sequencing of amplified single copy DNA. Nature (Lond) 330: 384–386

Wood WI, Capon DJ, Simonsen CC, Eaton DL, Gitschier J, Keyt B, Seeburg PH, Smith DH, Hollingshead P, Wion KL, Delwart E, Tuddenham EGD, Vehar GA, Lawn RM (1984) Expression of active human factor VIII from recombinant DNA clones. Nature (Lond) 312: 330–337

Wulff K (1983) Luminometry. In: Bergmeyer HU, Bergmeyer J, Graßl M (eds) Methods enzym analysis, 3rd edn, vol. 1. Verlag Chemie, Weinheim, pp 340–368

Wyman AR, White R (1980) A highly polymorphous locus in human DNA. Proc Natl Acad Sci USA 77: 6754–6758

Yanish-Perron C, Vieira J, Messing J (1985) Improved M13 phage cloning vectors and host strains: nucleotide sequences of the M13mp18 and pUC19 vectors. Gene (Amst) 33: 103–119

Yehle CO, Patterson WL, Boguslawski SJ, Albarella JP, Yip KF, Carrico RJ (1987) A solution hybridization assay for ribosomal RNA from bacteria using biotinylated DNA probes and enzyme-labeled antibody to DNA:RNA. Mol Cell Probes 1: 177–193

Yolken RH (1988) Nucleic acids or immunoglobulins: which are the molecular probes for the future? Mol Cell Probes 2: 87–96

Clues to the Organization of DNA Repair Systems Gained from Studies of Intragenomic Repair Heterogeneity

C. A. SMITH[1] and I. MELLON[1,2]

Contents

Abbreviation

4NQO, 4-nitroquinoline 1-oxide; ADA, adenosine deaminase; BrUra, 5-Bromo-2'-deoxyuracil; CHO, Chinese hamster ovary; CS, Cockayne's syndrome; DHFR, dihydrofolate reductase; HPRT, hypoxanthine-guanine phosphoribosyl transferase; kb, kilobases; MHC, major histocompatibility; MT, metallothionein; PD, cyclobutyl pyrimidine dimer(s); TEV, T4 endonuclease V; XP, xeroderma pigmentosum.

1 Introduction

The investigation of DNA repair processes in mammalian cells has largely been modeled upon results obtained with prokaryotes, most notably *E. coli*. The general aspects of the organization and functioning of a number of repair systems

[1] Department of Biological Sciences, Stanford University, Stanford, California 94305-5020, USA
[2] Present address: Department of Pathology, Markey Cancer Center, University of Kentucky, Lexington, Kentucky, 40536-0093, USA

in mammalian cells do indeed appear to be rather similar to those in prokaryotes (Friedberg 1984). Both utilize: (1) a broad-spectrum excision repair system that recognizes and removes a variety of bulky, DNA-distorting lesions; (2) many individual glycosylases that specifically remove a variety of altered or damaged bases to produce repairable abasic sites; and (3) systems for directly reversing certain lesions like 0-6-alkyl guanine. Although the strategies used may not be identical, both eukaryotes and prokaryotes have been shown to repair mismatches and double-strand breaks, and to possess tolerance mechanisms to facilitate the replication of DNA containing lesions (sometimes termed postreplication repair).

The majority of studies with mammalian systems have used cultured cells, usually those that display rapid growth, to allow the requisite genetic and biochemical manipulation. However, the fundamental differences between mammalian cells and bacteria must be taken into account when considering aspects of repair systems such as their regulation. These differences include the germline/somatic cell distinction and differentiation of function in multicellular organisms, their much larger and less informationally "efficient" genomes, and the great complexity of the DNA packaging system (chromatin) they use. Each bacterial cell represents an entire organism; loss of viability and mutagenesis, usually considered to be the biological end points of DNA damage that are ameliorated by repair systems, apply directly and uniformly. For multicellular organisms, the effects of damage could vary widely among different cell types. Only in the germline cells would mutations be heritable from the species standpoint, and the requirements for repair systems in rapidly dividing somatic cells might be very different that those of terminally differentiated cells. For any given somatic cell, most of its large genome contains information unnecessary for proper cell function, some of which must be kept in a repressed state. Finally, the various states of chromatin can modulate the introduction of certain types of damage into DNA and, as the proper substrate for DNA repair systems, different chromatin states might play a large role in the kinetics or efficiency of DNA repair.

To approach the issue of how repair systems might be influenced by different functional and structural aspects of the mammalian genome, considerable effort has been devoted in the past several years to examining repair in particular genomic components. One approach is to use traditional methods to examine repair or adduct frequency in DNA fractionated by some general procedure thought to be correlated to function or structure, e.g. separating active and inactive chromatin fractions according to some solubility property, or isolating highly repetitive DNA species. A second approach is to study repair in specific DNA sequences, such as genes and their flanking regions, whose activity or chromatin organization can be analyzed and, in some cases, experimentally manipulated.

With this second approach, a number of interesting and provocative results have recently been obtained with cultured human and rodent cells. To date, much of the work in this area has originated from investigators in the laboratory of P. Hanawalt, who have developed and refined methodology for these approaches. However, the field is now rapidly expanding with the entry of additional investigators, many of whom presented results at a recent international symposium (summarized in Smith 1988a). The notable result obtained by studying specific sequences has been the demonstration of "preferential repair" of transcriptionally active DNA in UV-irradiated cells.

Although a number of reviews and descriptions of the published work in this field have appeared (Bohr 1987; Bohr and Hanawalt 1988; Bohr and Wasserman 1988; Bohr et al. 1987; Hanawalt 1986; Hanawalt 1987) most of these concentrated on the work of the particular author(s) involved, and have not described in detail the impact of other types of studies on interpretation of results. Enough new material, much of it not yet published, is now available to us to warrant the present effort. Our goal is to examine whether current knowledge about repair heterogeneity can be used to develop a consistent model for the organization and control of repair systems.

2 Preferential Repair of Pyrimidine Dimers

2.1 The DHFR Locus

The most intensively studied specific sequences with respect to DNA repair are those in and around the gene coding for dihydrofolate reductase (DHFR) in both human and Chinese hamster ovary (CHO) cells. The methodology for measuring the frequency of cyclobutane-type pyrimidine dimers (PD) was originally developed using a line of CHO cells in which the locus had been amplified about 50-fold. This had facilitated the cloning and characterization of the gene and its surrounding sequences by other investigators, and the increased copy number was invaluable in prototype studies because of the increased sensitivity it afforded.

The methodology developed is simple in concept. Isolated genomic DNA is incubated with a restriction nuclease that cleaves DNA at known locations in and around a gene of interest. Equal portions of the DNA sample are either treated or mock-treated with an enzyme, T4 endonuclease V (TEV), that makes a single-strand scission specifically at the site of each PD. The DNA is denatured and electrophoresed in alkaline agarose gels to separate DNA molecules according to size, and then transferred to a support membrane. It is then hybridized with ^{32}P-labeled genomic or cDNA probe specific for a particular genomic sequence, usually resident on a single restriction fragment. With increasing frequency of PD, more and more of the fragments are cut, and the amount of hybridization at the position of unit length fragments is correspondingly diminished. The amount of hybridization can be measured by scintillation counting of excised bands or by scanning densitometry of autoradiograms; the latter method allows multiple determinations from a single membrane by repeatedly hybridizing with different probes. Comparison of the amount of probe hybridized to the bands from samples treated and untreated with TEV allows calculation of the fraction of fragments containing no lesions, termed the zero class. Assuming that PD formation and removal by repair systems is random over the entire collection of fragments, this value can be used with the Poisson equation to calculate the actual frequency of

PD in the fragment, and thereby measure its repair. In practice, a density labeling procedure is usually used to exclude replicated DNA from the analysis. The nature of the probes and fragments available for a particular sequence are important factors in using this method, which we refer to as Southern analysis. The method has been described in detail by Bohr and Okumoto (1988) as well as in many references cited below; further discussion of its use may be found in Smith and Hanawalt (Biotechniques, in preparation). A similar method was apparently developed independently by M. Lieberman and co-workers for studying human ribosomal genes; it was mentioned in a meeting abstract (Rajagopalan et al. 1984) but has yet to be fully described. The same general strategy was also used by Nose and Nikaido (1984) to measure frequencies of breaks and alkaline labile sites in specific sequences in human cells treated with X-rays and some carcinogens.

Our initial results with CHO cells were striking (Bohr et al. 1985). It was found that PD were very efficiently removed from a 14-kb fragment that comprises approximately the 5' half of the DHFR gene. Even at considerable doses, 60–70% of the PD appeared to be removed from the fragment in 24 h. In contrast, measurements of the removal of TEV-sensitive sites from the genome as a whole indicated very inefficient repair (about 15% removed in 24 h), in agreement with previous results with a number of rodent cell lines. Finally, repair was barely detectable in a DNA fragment thought to be located some 30–50 kb 5' to the DHFR gene, but still on the stretch of DNA amplified in these cells. From this, it was argued that the efficient repair in the gene was unrelated to the amplification, a point later confirmed by analysis of the parental cells in which amplification had not occurred (Bohr et al. 1986a) and by studies with other genes, discussed below.

These results suggested a resolution of the so-called rodent cell paradox. It had been well established that rodent cells in culture remove only a small fraction of PD, while at the same time exhibiting similar UV resistance to cultured human cells, which remove nearly all the PD from their genomes. For cells from both species, however, a functional bulky adduct excision-repair system is necessary for UV resistance, indicating that the limited repair in rodent cells was in fact important for survival. The paradox could now be explained by generalizing the proficient repair in the essential DHFR gene to all other vital sequences in the genome, and attributing the primary role of excision repair in UV resistance to removing blocks to transcription in such sequences. The ability of irradiated rodent cells to replicate DNA containing persisting damage could be ascribed to efficient tolerance mechanisms. (An alternative solution to the paradox, involving the role of another photolesion has also been proposed and will be discussed in a later section.) To account for the restriction of repair to vital sequences such as active housekeeping genes, it was proposed that chromatin structure of such DNA is more accessible to repair systems than the chromatin in other sequences. Some activity for rendering lesions in the silent DNA accessible to repair might be missing or have been lost during cell culture.

This heterogeneity in repair suggested as well that additional biological end points, such as mutagenesis, need not correlate with overall repair efficiency, and might vary according to the activity of the sequence involved. By inference, repair

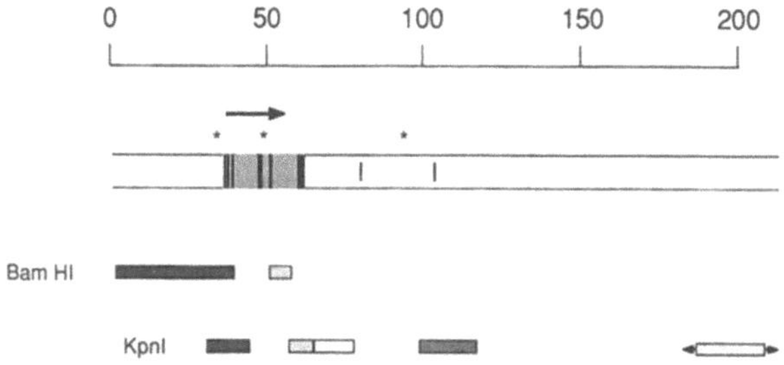

Fig. 1. Map of the region around the DHFR gene in CHO cells. The gene is depicted as *dark lines* (exons) and *shaded areas* (introns) with the direction of transcription indicated by the *arrow*. Locations of nuclear matrix attachments regions are indicated by *asterisks*. The two *vertical* bars downstream from DHFR represent the locations of initiation sites for DNA replication. The locations of restriction fragments used for analysis of repair are indicated by the *horizontal* boxes. The *arrows* on one fragment indicate uncertainty as to its precise location. The scale is in kilobases (kb)

heterogeneity in whole organisms could increase rates of mutagenesis in unexpressed genes, with potential implications for carcinogenesis.

A great deal of work has been applied to a more detailed analysis of repair in the sequences surrounding the DHFR gene in CHO cells (Fig. 1), facilitated by the cloning and characterization of about 275 kb in this region by J. Hamlin and collaborators (e.g., Montoya-Zavala and Hamlin 1985). The aim of these repair studies was to map the boundaries of the DNA stretch subject to efficient repair, and to help assess the role of transcription itself in the process. A number of interesting results have emerged.

1. A 14-kb fragment lying just 3' to the transcription unit has been consistently observed to be very inefficiently repaired in dividing cells (Bohr and Hanawalt 1986; Bohr et al. 1986b), and in nondividing cells even after 48 h (unpublished, see Bohr and Hanawalt 1988; Smith 1987), and in cells containing hypomethylated DNA (Ho et al. 1989). Interestingly, this fragment occupies the region between the end of the DHFR transcription unit and a segment of DNA that has been characterized as a replication origin region. This segment contains two initiation regions and a matrix attachment site (Dijkwel and Hamlin 1988).
2. A 30-kb fragment that contains the 5' third of DHFR, but extends about 20 kb in the 5' direction, is repaired nearly as well as the fragment that represents the 5' half of the gene. A divergent transcription unit lying 5' to the DHFR gene has been identified in CHO cells (Mitchell et al. 1986), and could account for the efficient repair in this fragment.
3. Fragments in the 3' half of the DHFR gene appear to be less well repaired than the fragment comprising the 5' half. This has been observed in two different investigations. Unfortunately the fragments available for study are rather small. They have been studied only in one cell line, which contains multiple copies of the locus, and after irradiation at 20 J/m^2, which places considerable damage in 5' portions of the gene. It is perhaps relevant that a matrix attachment sequence has been identified near the 3' end of the profi-

ciently repaired 14-kb KpnI fragment (Kas and Chasin 1987). Interestingly, more efficient repair has been observed for one of these fragments in cells containing hypomethylated DNA (Ho et al. 1989). A number of explanations for the low levels of repair can be put forward, but since proficient repair has been observed in 3' fragments of some other genes, the general relevance of these observations is unclear.

4. With the exception of the fragment just downstream from DHFR, no other fragment in the region has been found to exhibit consistently the poor repair characteristic of the genome overall. A 25-kb fragment centered about 60 kb downstream from DHFR exhibits about 55–65% removal of dimers in 24 h (Ho et al. 1989). This fragment is about as distant from the replication initiation region as is DHFR, but on the other side of it.
5. The 22-kb fragment we previously reported to be poorly repaired in both amplified and unamplified cells (Bohr et al. 1985; 1986a) has, in some subsequent experiments, shown 30–40% dimer removal (Bohr et al. 1986b; Ho et al. 1989). In the normal cell this fragment lies about 160 kb downstream from DHFR; in amplified cells it will also be found upstream of the gene because of the tandemly repeated nature of amplicons. Its exact position depends upon the location of the breakpoint of the amplicon, 5' to the DHFR gene. The amplified line we used for repair analysis has not been characterized in this regard. Looney et al. (1988) found widely varying sizes of amplicons in five different Chinese hamster lines, so it is possible that this fragment in our cells is quite distant from the 5' side of the DHFR gene. In cells with hypomethylated DNA, repair in this extragenic fragment was increased to the levels found in DHFR. This suggests the presence of a transcription unit on this fragment; perhaps the level of repair observed and its variability reflect variable amounts of transcription under different conditions.

It would be instructive to examine a fragment located just 3' to the unidentified gene that shares the promoter region with DHFR, to see if that DNA is also poorly repaired, but the gene is not yet adequately characterized. Without further identification of other putative transcription units in this region and characterization of their activities, we cannot determine from these studies whether the generally proficient repair in the sequences analyzed in the DHFR "domain" is the result of gene activity, or stems from some general property of the chromatin structure in this area. The deficient repair in the sequence 3' to the gene itself could be atypical, possibly due to some special protein associations related to its position near a replication origin.

It could be argued that the preferential repair of the DHFR gene in CHO cells is merely the result of the poor removal of dimers from the overall genome in these cells. For human cells, which exhibit proficient removal of dimers, we theorized that preferential repair might be reflected in the *rate* of removal of dimers from genes. Repair was examined in the DHFR locus in a human cell line, also carrying an intrachromosomal amplification of the region (Mellon et al. 1986). We observed 65–80% removal of dimers in only 4 h following doses of 5 or 10 J/m^2, both from a 20-kb KpnI fragment comprising most of the gene, and from a

25-kb KpnI fragment situated just 5' to DHFR (Fig. 2). A separate analysis of repair in the genome overall indicated only about 25% removal in this period, followed by 65–90% removal by 24–48 h. To confirm the apparent difference in rates, the method used for analysis of total DNA (Van Zeeland et al. 1981) was extended to allow simultaneous assessment of repair in the DHFR region. At various times after irradiation, permeabilized cells were treated with TEV and their DNA sedimented on alkaline sucrose gradients. By probing the fractionated gradients, it was possible to show that after 4 and 8 h, the inter-dimer distance was substantially greater in the DNA molecules that hybridized to the probe for DHFR than for the mass-labeled DNA or DNA that hybridized to the nontranscribed alpha DNA. This technique allows only approximate calculation of the amount of repair, but the values obtained were consistent with results obtained by Southern analysis.

These results indicated that preferential repair also occurs in the DHFR locus in human cells. The rapid repair observed in the sequence 5' to the gene could result from the presence of a divergent transcription unit in this locus, analogous to the one in CHO cells (Mitchell et al. 1986). A number of possibilities were suggested to explain the rapid repair of active genes. Additional processing required to make inactive DNA accessible to repair could decrease its rate of repair, fundamentally different repair systems could operate on active and inactive DNA, or the blockage of transcription by a lesion might make it a more attractive substrate.

Rapid repair of active genes had been suggested by Mayne and Lehmann (1982), who observed that the resumption of RNA synthesis in UV-irradiated human cells is more rapid than the overall kinetics of excision repair of PD. Cockayne's syndrome (CS) cells, in which the rapid resumption of RNA synthesis was found to be deficient, are hypersensitive to UV but show no clear defect in overall excision repair. It appeared that a significant biological role for preferential (rapid) repair could be established by demonstrating its deficiency in Cockayne's syndrome cells. Further evidence of the importance of rapid repair is provided by the demonstration that UV-irradiated cells exhibit greatly enhanced killing when incubated with an inhibitor of DNA repair synthesis (aphidicolin) only during the first few hours after irradiation and not at later times (Keyse and Tyrrell 1987). Since dimers have been shown to be effective transcription-blocking lesions, these observations suggested that survival after UV depends to a large extent on the

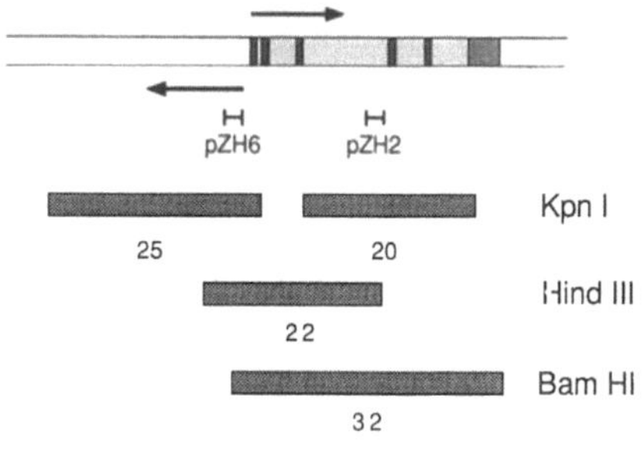

Fig. 2. Map of the region around the DHFR gene in human cells. Details are similar to those for Fig. 1, with the exception that the sizes of fragments are listed below them in kilobases and the locations of the two fragments used for synthesis of strand-specific probes are shown directly below the representation of the DNA

cell's ability to maintain some minimum level of expression of certain critical genes. This had been previously proposed by Kantor et al. (1977) and Konze-Thomas et al. (1982). However, it was puzzling that the rate of repair observed in the CHO DHFR gene appeared to resemble the relatively slow rate of overall repair in human cells, rather than the rapid rate of repair of the human DHFR.

The experiments discussed thus far all used probes made by nicktranslation; both complementary strands of DNA were being analyzed simultaneously. Mellon et al. (1987) constructed plasmids from which RNA probes can be synthesized that hybridize specifically to one or the other strand of DNA and analyzed repair in the individual strands. They discovered that the preferential repair observed in the 14-kb fragment within the CHO DHFR gene was confined to the transcribed strand. Dimer removal in the strand was indeed rapid, and nearly complete in 8 h, whereas removal in the nontranscribed strand varied between 5–12% over the entire 24 h period. Similar results have been obtained with a CHO line not containing amplified DHFR (Mellon et al. 1988). In human cells, the transcribed strand of DHFR was repaired with the same kinetics observed for this strand in the CHO gene, while repair in the nontranscribed strand was slower and less complete, resembling the repair observed for the human genome as a whole. It was in fact the differences in repair in their *nontranscribed* strands that had previously given the appearance of different kinetics for repair in the two different DHFR genes.

The fragment 5' to the human DHFR was also analyzed in this way and similar differences between repair in the two strands were observed. In this fragment, the rapidly repaired strand was contiguous with the nontranscribed strand of the DHFR gene. This is consistent with the hypotheses that (1) rapid repair is a property of the transcribed strand in a gene, and (2) the 5' flanking fragment contains a gene whose polarity is opposite to DHFR, as suggested by sequence homology to the same regions in the DHFR locus of mouse and CHO cells (Mitchell et al. 1986).

These results have several important implications for studying repair in specific sequences and for considering the mechanism and consequences of preferential repair. Obviously, when the two strands of a sequence are repaired at different rates, but analyzed together, erroneous conclusions can be drawn. Except in cases where lesion levels are low, the Poisson analysis provides an estimate of repair that exceeds the average of the actual values for the separate strands. The effects of preferential repair of only one strand in a sequence on its replication and mutagenesis are harder to predict than are effects of repair of both strands. Models for the mechanism involved must be more detailed. Unfortunately, thus far nearly all of the data now available for other genes and adducts were obtained by analyzing both strands together.

Regardless of how general strand specificity of repair turns out to be, it is important to consider models for its mechanism. It is unlikely that the transcription-template strand of an active gene is simply more accessible to repair enzymes than is the other strand. It seems much more likely that the bias in repair is directly related to different effects of DNA lesions in the respective strands on the transcription process itself. UV photoproducts in DNA are known to cause termination of transcription both in vivo and in vitro, resulting in the buildup of truncated transcripts. It seems reasonable to suppose that only lesions on the template strand have this effect, as has been shown in vitro for psoralen monoadducts (see below). An enhancement of repair activity at sites where transcription is blocked would at once explain both preferential repair in active genes and its strand bias.

Lesions blocking transcription could be attractive substrates for repair for a number of reasons. Transcription complexes are associated with the nuclear matrix; blocking lesions might thus be held stably at this structure, perhaps thereby increasing their accessibility to repair proteins. Blockage could also result in some relatively long-range alteration in the chromatin structure which facilitates the association of repair proteins scanning the DNA for damage near the site of the lesion. The probability that the repair system recognizes the dimer as DNA damage may be increased by the presence of the transcription complex or remnants of it. The configuration of a lesion at the site of blockage may resemble the one brought about by preincision repair activities, thereby bypassing the need for some preincision processing. For example, if DNA unwinding at the site of a lesion is requisite for the binding of some repair protein, the unwinding at the transcription complex may circumvent the need for some specific component(s) of the repair system. This possibility is attractive becauce a deficiency in such a preincision step could affect repair in active and silent DNA to different extents. The recent demonstration of more rapid repair in the transcribed strand of the lactose operon only when it is induced in UV-irradiated *E. coli* (Mellon and Hanawalt 1989) suggests that enhanced repair at blocked transcription complexes may be a general phenomenon, unrelated to eukaryotic chromatin structure.

At present we have no evidence that transcription itself is required for strand-specific repair in mammalian cells; it is possible that there is a distinct repair system for scanning genes, which interacts with the DNA in much the same way as the transcription complex. A repair protein or complex might associate with DNA at the same chromatin elements used for entry of RNA polymerase, and translocate in the same direction along the template strand. Such a system could operate only in domains containing a gene having the proper chromatin configuration, but once initiated, might continue past transcription termination signals, thereby functioning in regions not normally transcribed. Other signals, such as matrix attachment sites or replication origins, could, however, terminate its operation. For repair to be strand specific, the mechanism for recognition of damage would need to be sensitive to alterations in unwound DNA.

2.2 Preferential Repair in Other Housekeeping Genes

The essential features of preferential repair have been confirmed in several other genes, both in rodent and human cells. In a 21-kb fragment encompassing most of the 25-kb gene encoding 3-hydroxy-3-methylglutaryl coenzyme A (HMG CoA) reductase in CHO cells, dimers were removed to about the same extent as in DHFR (Smith 1987). (Strand specificitiy of repair has not yet been examined in this gene.) In the Chinese hamster line V79, proficient repair has been observed in two fragments that span most of the hypoxanthine guanine phosphoribosyl transferase (HPRT) gene (L. Mullenders, pers. comm.) In mouse 3T3 cells, Madhani et al. (1986) observed efficient repair of the active c-*abl* gene in both growing and nondividing cells. However, very little repair was observed in 15- and 6-kb fragments that contain portions of the mouse c-*mos* proto-oncogene, whose transcription has not been detected in these cultured cells. However, no precise map of the c-*mos* gene in these cells is available to establish the positions of the fragments analyzed with respect to this transcription unit.

In human fibroblasts, preferential repair has been observed in a large fragment of the c-*abl* gene (I. Mellon, unpublished observations). The preferential repair of the 32-kb adenosine deaminase (ADA) gene was reported by Mayne et al. (1988). In confluent normal fibroblasts, the rate of dimer removal from an 18-kb EcoRI fragment that extends about 8 kb into the 3' flanking region of ADA was similar to that reported by Mellon et al. (1986) for DHFR in immortal, rapidly growing human cells, and the slower repair of the overall genome was also similar in fibroblasts and immortal cells. Preferential repair has also been observed in a different fragment, lying entirely within the ADA gene (L. Mullenders et al., pers. comm.). Preliminary results appear to show rapid repair of both strands of the 3' fragment of the ADA gene. However, there is evidence (R. Kellums, pers. comm.) for a *convergent* transcript in this region; thus both strands in this fragment may be part of transcription units. As a control, Mayne et al. (1988) analyzed repair in a 14-kb EcoRI fragment that lies in an unexpressed region of the X chromosome, termed the 754 locus. This fragment was repaired slowly, with the same kinetics as the overall genome. Kantor et al. (submitted) measured repair in a 14-kb fragment containing the 3.5 kb expressed beta-actin gene in nondividing human fibroblasts, probing both strands simultaneously. Very rapid repair in this fragment was observed; greater than 50% removal occurred in 4 h. However, in contrast to the results of Mayne et al. (1988) in these cells, considerably less repair was observed in the silent 754 locus than in the overall genome.

2.3 Ribosomal Genes

Although these may be also be considered housekeeping genes, they differ in many respects from the genes encoding enzymes discussed to this point. The ribosomal genes are organized into clusters of tandemly repeating units present in multiple copies on several chromosomes. Each unit is composed of a spacer sequence and a sequence that is transcribed into a precursor RNA molecule, which is subsequently processed to form the three mature ribosomal RNAs. Transcription is localized in the nucleolus, and is mediated by a distinct RNA polymerase, polymerase I. Ribosomal DNA also has an abnormally high G+C content, a property that has been exploited for isolation of DNA rich in ribosomal sequences. This affects their sensitivit to damage formation; Rajagopalan et al. (1984) observed only about 25% as many TEV-sensitive sites in human ribosomal sequences as in bulk DNA, even when DNA was irradiated in vitro. They reported that repair could be detected in these sequences 24 h after UV, but did not indicate the amount of repair. Cohn and Lieberman (1984) mentioned, but did not document, that they observed from 25–40% as many repaired sites in ribosomal sequences as in the bulk of the DNA, using a method similar in principle to the method devised by Leadon, described below. Taken together, these two undocumented statements suggested that repair in ribosomal genes resembled that in the bulk of the DNA.

To determine the kinetics and strand specificity of repair in these actively transcribed genes, removal of TEV was measured in a 13.5-kb fragment that encompassed most of the transcription unit in human fibroblasts irradiated with 20 J/m^2 UV (I. Mellon, unpublished). The initial frequency of TEV measured was about half that usually observed for other genes. Neither rapid nor strand-specific repair was observed. After 24 h, only about 40% of the sites were removed, determined by using probes either for both strands or the transcribed strand specifically.

To interpret these results, one must take into consideration the possibility that under the conditions used, only a fraction of the many sequences in the cell may be active, and important features of repair in the active fraction may be masked by characteristics of repair in the remainder of the genes. (We should note that functional hemizygosity in the case of genes not occurring in multiple copies might, in some cases, similarly influence the interpretation of repair analysis). It is also possible that the organization of the active genes into the nucleolus or some aspect of the mechanism of transcription by RNA polymerase I affects repair. Analysis of isolated nucleolar DNA may allow resolution of this question. The redundancy of these genes, especially when combined with their low dimer yield, may lessen the need for rapid and complete repair.

3 Preferential Repair: Relationship to Transcription

The demonstration of preferential repair in housekeeping genes and its absence in some nontranscribed sequences suggested that it is related to gene activity. This is consistent with a plausible biological role: namely, to promote the rapid recovery of the activity of genes whose products are necessary for cell function. The finding that preferential repair is confined to the transcribed strands of DHFR genes further supported some direct relation to transcription.

In addition to comparing repair in active and inactive sequences in the same cell, various approaches, which involve examining the same gene under different conditions, have been taken in an attempt to define the relationship between repair and transcription in more detail. Gene activity can be modulated in a single cell system with inhibitors or natural effectors, such as gene-specific inducers or suppressors. In such cases, it is usually the presumption that the rate of transcription of the gene is affected directly by trans-acting molecules, and that the chromatin structure of the gene is in some "permissive" state even when activity is low. A number of cell systems can be induced to undergo differentiation in culture; in such cases it should be possible to compare the repair in genes after they have undergone fundamental changes in potential for transcription; presumably these are related to chromatin structure and/or DNA methylation. In these systems, some tissue-specific genes can become activated during differentiation, while other genes no longer required for cell function may be suppressed. However, because the overall state of the cell changes, it is necessary to consider the possibility that regulation of repair may also be changed during the differentiation process. Finally, repair can be assayed in a tissue-specific gene in different activity states by different cultured cell lines, each of which retain gene expression characteristics of the different tissues from which they were derived.

3.1 Inducible Genes

Since inducible genes often maintain a basal level of expression, it is useful to consider how the actual rates of transcription might be expected to influence repair. Using the model that the blockage of transcription by a lesion somehow elicits preferential repair, we can make some detailed speculations. Increasing the rate of transcription initiation in a gene containing a lesion should increase the probability that the lesion would be in some favored configuration. The "lifetime" of such a configuration would then determine the relationship of transcription initiation rate to repair rate. If this lifetime were very long, e.g., if the blocked transcription complex were very stable and if it were to be the substrate for repair, even a seemingly low rate of initiation, say once an hour, might be sufficient to effect substantial repair during the first 2 or 4 h after damage, usually the shortest time intervals examined. The observation of preferential repair might in fact be a very sensitive indicator of even a low level of transcription. A 10–20-fold increase in transcription rate, brought about by some inducer, would be considered a large effect on that process, but might not have a correspondingly large effect on repair rate. On the other hand, if a blocked transcription apparatus should readily disassemble, a favored configuration at the lesion would have a relatively short lifetime. This could make repair exquisitely dependent on the rate at which transcription is initiated. More elaborate models and situations can easily be envisioned, which lead to threshold effects and non-linear dependencies. In most cases we have little or no information about actual rates of transcription, only about comparative rates.

We chose to study repair in the HMG CoA reductase gene in CHO cells because the enzyme activity in these cells can be manipulated by altering the composition of the cell culture medium. In medium supplemented with lipid-poor serum, cells are deprived of their usual source of cholesterol, and enzyme activity can increase over 20-fold. Supplementing the normal medium with sterols and high concentrations of mevalonate (the product of HMG CoA reductase) causes a reduction in enzyme activity. Although regulation of the protein has been shown to occur at the level of transcription (Lusky et al. 1983) regulation via effects on translation and protein degradation have also been demonstrated (Nakanishi et al. 1988).

We have not observed any significant reduction in the total amount of repair in this gene when cells were grown under suppressive conditions (C. A. Smith, unpublished observations), although some indications of a slower initial rate were obtained. The basal level of expression of the gene may be high enough to ensure its repair. These experiments utilized nick-translated probes, and accurate measurement of different rates of repair might require examination specifically of the transcribed strand.

An inducible system that has received considerable study is the metallothionein (MT) gene family (Hamer 1986). Unlike the housekeeping genes we have so far discussed, these are small genes, of the order of only 1 kb. Thus the fraction of genes containing damage is small (20%) even at a dose of 20 J/m^2. Analysis has therefore been applied to larger DNA fragments containing the genes.

Okumoto and Bohr (1987) used CHO cells that are more resistant to the toxic effects of cadmium by virtue of an amplification of the genes MT I and MT II. The study focused an 6- and 18-kb restriction fragments that contain the gene coding for MT I; the protein concentration has been reported to increase 1000-fold in a few hours in cells treated with $ZnCl_2$. In the absence of treatment with Zn, repair measured at 8 or 24 h after 10 or 20 J/m^2 mostly fell in the range of 14–38%. These are consistent with values previously reported for nontranscribed sequences or the bulk of the DNA in rodent cells. When a 2-hour treatment with $ZnCl_2$ preceded irradiation, repair increased by roughly 50% in each fragment. When irradiation followed immediately after $ZnCl_2$ addition, repair in the 6-kb fragment was 44% after 10 J/m^2, both at 8 and 24 h. This increase appeared to be significant because the control values were atypically low (6 and 9%). No effect of $ZnCl_2$ was observed for repair in the DHFR gene, or on overall repair synthesis in the cells.

These results are both puzzling and difficult to interpret. It should be emphasized that obtaining data with the precision necessary to distinguish small differences in repair is not simple, even with amplified genes. The situation is further compounded by the lack of characterization of the locus. Complete and rapid repair in the MT gene alone would not account even for the increase observed for the small fragment, since the gene occupies less than 20% of it. The locations of the MT genes on the fragments are unknown, and there is apparently no data available on whether other transcription units might be located on them. The notion that a large "repair domain" is activated by induction of MT genes could help explain these results, but further study, with strand-specific genomic probes, and fine-structure analysis of the region involved, would be needed to substantiate it.

Repair was also measured in a fragment containing MT II, a gene reported to be constitutively expressed at a much higher level than MT I in this cell line (Crawford et al. 1985). Significantly more repair in this gene than in MT I might therefore have been expected in the absence of induction, but this was not reported. The possibility that UV itself induces MT genes in these cells, as has been reported for V79 cells (Fornace et al. 1988), further complicates the interpretation of these experiments.

Leadon and Snowden (1988) analyzed repair in fragments containing several members of the large MT gene family in human cells. They used a technique (Leadon 1988) in which excision repair itself tags the DNA at the sites of repair with the base analogue BrUra (5-bromo-2'-deoxyuracil), after which DNA restriction fragments containing these repair patches are physically separated from the remainder of the DNA using a BrUra-specific monoclonal antibody. Southern analysis of these populations can then be used to measure the fraction of fragments containing a particular sequence that have been subject to a repair event. This can then be compared to data for other fragments, or to repair in the genome overall, given by the proportion of antibody-bound, mass-labeled DNA.

This method differs in important respects from the endonuclease method. It provides comparative estimates of repair, but not actual lesion frequencies, and measures not a single specific lesion, but the sum of all repair events that produce the repair patches containing BrUra. However, it is likely that a single repair event will place an entire fragment into the "repaired" class, regardless of the lesion frequency on the fragment (Leadon 1986), whereas the endonuclease method requires all lesions to be removed for a fragment to fall into the repaired class. Because the "background" value for the "repaired" class is essentially zero, even small amounts of repair are detectable, and differences in repair, even at low absolute levels, are easily observed. Small amounts of repair are not easily detected or compared with the endonuclease method because it requires the accurate measurement of small differences between two relatively large numbers, the zero classes. Fragments of different size can be simultaneously analyzed with the antibody method because it does not entail deliberate breakage of DNA. (The deliberate breakage used in the endonuclease method produces hybridization smears in the autoradiograms below the position of the largest band that hybridizes to the probe. These smears confound the analysis of smaller fragments.) This feature of the method was exploited in the study of MT genes in human cells; four different fragments with different levels of expression and sensitivity to induction were compared at the same time.

At early times after UV, repair in fragments carrying the unexpressed I_B and II_B genes resembled repair in the genome overall. About twice as much repair was observed in two other fragments, one containing MT II_A and one containing both MT I_A and I_E, all of which are constitutively expressed at some basal level. Inducing treatments (cadmium or steroids) increased repair about another factor of two for genes responsive to the treatments, but not for the others. Thus, repair in fragments carrying unexpressed genes was not affected by treatment with either inducer. With cadmium, which induces all three of the expressed genes to high levels, repair was increased in the fragments containing them. Steroid treatment further increased repair only in the fragment carrying II_A, the only gene in this collection responsive to it. At later times after irradiation, the amounts of repair increased in all fragments and was roughly correlated with their sizes. These results are consistent with the notion that the rate of repair of these MT genes is correlated with their rate of transcription, but that eventually most of the UV-induced lesions are removed from all of them.

Southern analysis has been used to measure repair in fragments containing the mouse beta globin genes in an erythroleukemic cell line in which transcription of the genes can be greatly induced by treatment with dimethyl sulfoxide (Haqq and Smith 1987). These are small (~2 kb) and part of a gene family. In the mouse, the beta-major and beta-minor genes are located about 15 kb from each other and reside on distinct EcoRI fragments of 7 and 14 kb, respectively, both of which hybridize to a probe containing the 5' portion of beta-major. Using Southern analysis for measurement of TEV-sensitive sites, we were unable to detect repair in either fragment in uninduced cells 24 h after irradiation with 20 J/m^2. In induced cells shown to be producing hemoglobin by a staining assay, no repair was detected in the larger fragment, but about 30% repair in 24 h was observed for the smaller fragment. This is the expected result if proficient repair in these fragments is confined to the 2-kb genes themselves, but we consider the results only suggestive, due to the magnitude of variation reported for repair in unexpressed sequences. A more firm conclusion is that inducing these genes does not greatly increase the repair in a large genomic domain containing them. As expected from the previous data concerning the inactive c-*mos* gene in mouse cells, no repair was observed in the fragments containing globin genes in a mouse fibroblast line, not expressing globins.

3.2 Comparisons of Genes in Different Cell Types

Another approach for correlating repair to gene activity is to study genes that are expressed in one cultured cell type but not in another. This offers the opportunity to compare genes whose activity and chromatin structure is determined by the developmental history of the tissue of origin. We took this approach in studying the genes of the major histocompatibility complex of the mouse. We focused on four restriction fragments that occupy most of a 75-kb stretch of the genome and contain the A_α gene, E_β gene, an unexpressed pseudogene containing some homology to E_β (termed E_{β_2} and the E_α gene (Fig. 3). The fragments range in size from 8-17 kb. In the case of the three genes, each is completely contained in its respective fragment and comprises at least 70% of it. Repair was measured 24 h after irradiation with 20 J/m^2, using probes that detect both strands simultaneously (Haqq and Smith 1987; Haqq and Smith, unpublished).

In the M-12 B cell line, which expresses these genes, 50-70% of the dimers were removed from the genes, but only 22% repair was observed in the fragment

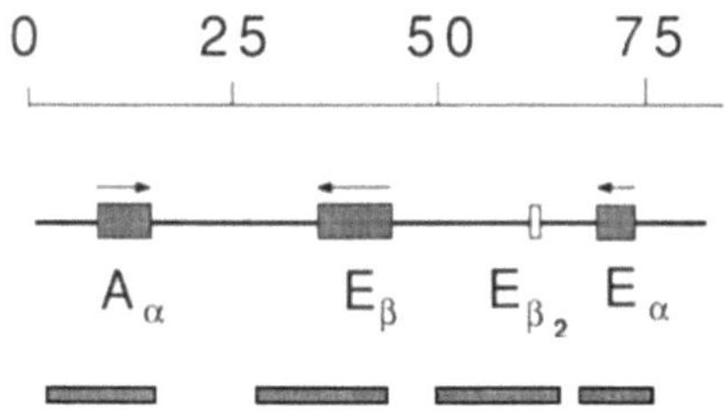

Fig. 3. Map of the region in mouse cells containing the MHC (major histocompatibility) genes analyzed for repair. Details are similar to those for Fig. 1

containing the E_{β_2} sequence. This fragment comprises about 80% of the DNA located between the E_β and E_α genes. The lack of repair in the silent region between two active genes suggests a strong correlation between repair and transcription in this case.

A plasmacytoma cell line was also examined, because it is derived from cells that had passed through the B-cell stage in situ and then presumably ceased expressing these genes. It had been shown by antibody staining that neither the I-A or I-E molecules (each composed of α and β chains of the relevant type) were present on the cell surface. Indeed, it was found that the repair in fragments containing E_α, A_α and the psuedogene were all in the 20% range. In 3T3 fibroblasts, no detectable repair was observed for E_α and A_α. Unexpectedly, proficient repair was observed in the E_β gene in both plasmacytoma and fibroblast cells, suggesting some low level of expression of this gene. However, this result was not confirmed in a recent experiment with a different fibroblast line (Boyajian and Smith, unpublished).

3.3 Differentiated Cells

Restricting repair to active genes would seem to be advantageous to terminally differentiated cells. With no need to replicate their DNA or protect genes needed for future cell generations, such cells could concentrate their repair resources on genes needed for metabolic and tissue-specific activities. It is conceivable that the lack of repair in the overall genome in rodent cells results from a premature regulatory event that normally takes place in differentiated cells. Previous studies of repair in such cells generally have shown a decline in overall repair capacity as cells become terminally differentiated, but could be questioned; the radiolabeling methods used for comparative analysis might be sensitive to different nucleotide kinase activities or precursor pools in the differentiated cells. Southern analysis of repair does not depend on radiolabeling and thus avoids this difficulty.

Kessler and Ben-Ishai (1988) studied a rat myogenic clone (L8E63) that can differentiate in culture to form myotubes when the cells are allowed to reach confluence and deprived of serum. They examined removal of TEV-sensitive sites in two different ~9-kb EcoRI fragments, each containing about half of the creatine kinase gene, whose expression should be induced in the differentiated cells. They were unable to observe significant repair in the gene in 24-h periods after UV-irradiation was given to rapidly growing cells, or to differentiating cells, either at a time just after the onset of expression of muscle-specific proteins or several days later. Similar results were obtained for rat fibroblasts, and preliminary data of a similar nature were mentioned for the more synchronously differentiating L8 cells. This result implied that repair in the differentiated cells is simply very limited. However, no positive control of a repair analysis in a housekeeping gene was available.

Two differentiating rat cell systems are currently being used in our laboratory. L. Ho is studying the L8 muscle cell system just mentioned and P. Gee is using a model nerve cell system, in which PC12 pheochromocytoma cells are induced to differentiate by addition of nerve growth factor to the culture medium.

In the undifferentiated L8 cells, several genes thought to be unexpressed do exhibit low levels of repair (0–20%) as expected if these cells behave like the other rodent cells previously studied. These genes include the embryonic myosin heavy chain gene, which is induced upon differentiation. Two active genes, triosephosphate isomerase and pyruvate kinase, showed 35–40% removal of dimers in 24 h. While this is more proficient repair than observed in the silent genes, its extent and rate (from determinations made at shorter times) is considerably less than that observed in housekeeping genes previously studied in mouse or CHO cells.

In the differentiated L8 cells, repair in the myosin gene in the first 24 h after irradiation did not appear to be greater than that for undifferentiated cells, measured either immediately after the start of the differentiation process, or at later times. These results resemble those reported by Kessler and Ben-Ishai. However, when the repair period was extended, proficient repair was observed, and after several days it reached the 50–70% level characteristic of active genes in CHO and mouse cells. This slow rate of repair probably accounts for the negative results of Kessler and Ben-Ishai. Surprisingly, this slow but proficient repair in the differentiated cells was also observed for several inactive genes, chosen as negative controls for these studies because their activity was not expected to be induced in this system. It is of course possible that in the undifferentiated cells, these genes would be repaired given enough time. However, no significant increases in repair were observed at 48 h in growing cells. Such measurements were possible because these rat cells do not replicate their DNA as proficiently after UV as do CHO cells; thus appreciable amounts of unreplicated DNA are still available for analysis at 48 h. However, there is always the possibility that the DNA of poorly repairing cells is selectively analyzed in such an experiment. To carry out studies over 4–5 days would require suppressing normal DNA synthesis in some way, and has not yet been attempted.

Analysis of total RNA in these cells has confirmed the induction of the myosin gene in the differentiated cells, but induction of several control genes was not observed. It seemed possible that repair of the unexpressed genes might result from a general change in the repair characteristics of the differentiated cells leading to proficient repair over the entire genome. To test this, direct comparisons of the removal of TEV-sensitive sites were made by mixing differentially labeled genomic DNAs from cells in the different stages and analyzing them on sucrose gradients. Similar amounts of repair were observed for differentiated cells allowed 4 days to repair and undifferentiated cells allowed 24 h.

Unexpected results have also been obtained with the PC12 cells. These cells grow very slowly, doubling every 2–3 days. In none of the genes examined did repair levels exceed 20% in the first 24 h following irradiation. After 96 h, repair in the expressed pyruvate kinase gene reached only about 40%. In this system the gene for the growth associated protein (GAP-43) is of interest because it is induced from a basal level in the untreated cells to high levels by nerve growth factor. In untreated cells about 30% dimer removal from a fragment in this gene was observed 72 h after irradiation; this rose to about 50% under the same conditions in the differentiated cells.

Clearly, the expectation that these differentiated cells might exhibit repair characteristics similar to the other rodent cells we have studied, but they would have different, cell-specific spectra of proficiently repaired genes, has not been fulfilled in these systems. However, since repair, even in the undifferentiated cells used, exhibits features different from those in the rapidly growing fibroblast-derived lines usually studied, more information will be required from a variety of systems to obtain an adequate picture of the relevant features of regulation of repair in terminally differentiated cells.

4 Repair in Mutant Cells

4.1 Xeroderma Pigmentosum Group C Cells (XP-C)

One of the first reports of genomic heterogeneity in repair concerned xeroderma pigmentosum (XP) cells belonging to complementation group C. These cells are only partially deficient in removal of bulky adducts, and do carry out a limited amount of repair synthesis. Mansbridge and Hanawalt (1983) grew cells in ^{14}C-thymidine to label the DNA uniformly, then held them in quiescence to suppress DNA replication, irradiated them with UV and labeled the subsequent repair synthesis with tritiated thymidine. DNA was purified from the cells, treated with TEV and analyzed on alkaline sucrose gradients. The molecules containing the tritium-labeled repair tracts were on average considerably larger than the bulk of the DNA molecules, identified from the ^{14}C label. In repair-proficient cells, the molecular profiles were congruent. These results indicated that, although the number of repair events in XP-C cells is much smaller than in normal cells, they do not occur randomly in the genome, but rather in a clustered manner. This was interpreted to mean that only certain DNA "domains" were repaired in these cells, and the phenomenon was termed "domain-limited repair". It appeared that within the domains, the extent of repair approached that in normal cells. In subsequent investigations, Mullenders et al. (1984), Karentz and Cleaver (1986), Kantor and Player (1986), and Kantor and Elking (1988) all reproduced and further explored this phenomenon in a variety of different XP-C strains. It was not found in XP-A or XP-D cells.

Cleaver (1986) examined domain-limited repair in XP-C as a function of the growth state of the cell. He was unable to observe it in exponentially growing cells, and found that it was lost after confluent cells were stimulated to enter the cell cycle, and returned when they again reached confluence. He also reported (1987) that incubation in a concentration of aphidicolin (an inhibitor of DNA polymerases alpha and delta) that reduced repair synthesis by about 50% resulted in loss of domain-limited repair in nondividing cells. [However, Mullenders et al. (1984, 1986) did not observe changes in the repair characteristics they measure (see below) when cells were incubated in hydroxyurea and cytosine arabinoside, conditions which also inhibit these polymerases.] Interpretation of these experiments is complicated by the complex biochemical manipulations needed in these experiments to exclude labeling of DNA by normal synthesis and to counteract the effects of aphidicolin on the ligation of repair patches. They do suggest that differences in the state of the cells are important factors to consider in attempting to correlate the mode of repair in XP-C to biological end points.

At about the same time that domain-limited repair was reported, Mullenders et al. (1984) observed that repair synthesis in an XP-C cell strain appeared to be more closely associated with the nuclear matrix than the repair in normal cells. In these experiments, they also differentially labeled bulk DNA and repair tracts, but then progressively digested the DNA in salt-extracted nuclei with DNase I, and analyzed the products by sedimentation on neutral sucrose gradients. With no digestion, all the DNA sediments very rapidly, presumably due to its attachment to an intact nuclear matrix. With increasing extents of digestion, more and more slowly sedimenting free DNA molecules are released, until only a few percent of the DNA remains bound to the matrix. During the course of the digestion, the ratio of repair label to bulk DNA label in the rapidly sedimenting component from XP-C cells gradually increased to a final value of three- to four fold greater than the starting ratio. This result was not observed for normal cells or XP-D cells (Mullenders et al. 1986) treated in the same way, and a very slight effect was observed with an XP-A strain (XP8LO) that performs a moderate amount of repair synthesis. Domain-limited repair was not observed for these particular XP-A cells by Kantor and Player (1986). Additionally, a microscopic-autoradiographic method indicated a greater association of repair synthesis for XP-C cells than for normal cells with a structure thought to represent the matrix. This was done with exponentially growing cells, and necessarily excluded cells in S-phase due to interference by normal DNA replication. These qualitative observations appear to contradict Cleaver's finding of a lack of domain-limited repair in growing XP-C cells. Perhaps matrix association and domain-limited repair as operationally defined do not measure precisely the same processes.

Mullenders et al. (1983) had previously used the digestion method to determine whether repair synthesis itself is associated with the nuclear matrix, as appears to be the case for replication and transcription, and had concluded that it was not. Taking domain-limited repair in XP-C and the association of active genes with the matrix into account, they suggested that the association of repair synthesis in XP-C results from confinement of repair activity in these cells to transcriptionally active DNA. The model implied by this suggestion was that this DNA could be repaired because of its more open chromatin structure, but that XP-C cells lacked the ability to render the rest of the chromatin accessible to repair enzymes. Player and Kantor (1987) showed that DNA repair synthesis in XP-C cells was also more sensitive to endogenous nucleases, activated by incubating nuclei in certain buffers, than the bulk of the DNA. Since transcriptionally active DNA had previously been shown to be most sensitive to this digestion, these results supported the notion that XP-C cells retain a normal repair process that is associated with such DNA.

Similar suggestions were made by Bohr et al. (1985) to explain how CHO cells could efficiently repair the DHFR gene, while displaying low overall repair levels. They also proposed that this preferential repair of vital sequences was largely responsible for these cells' resistance to UV. The reported lack of repair in the DHFR gene of XP-C cells (Bohr et al. 1986a) provided at least one explanation of the difference in UV resistance between CHO and XP-C cells: repair in the latter apparently did not extend to all genes.

However, recent developments complicated that simple interpretation. Using Southern analysis, proficient repair has been observed in nondividing XP-C cells in fragments in the large ADA (adenosine deaminase) (Mayne et al. 1988) and DHFR genes (Mullenders, submitted). Kantor et al. (submitted) observed proficient repair in a 14-kb fragment containing the 3.5 kb β-actin gene and ~30% repair in the DHFR gene. The fragment in the silent 754 locus has exhibited very little repair. Although proficient, the repair in the β-actin gene was considerably slower in XP-C than in normal cells. Lower rates and extents can also be seen in data available for the other genes, so it seems likely that repair in XP-C domains does not exactly mimic that in normal cells.

In addition, evidence of a different sort for efficient repair of genes in XP-C has been obtained (Kantor and Barsalou 1988; Kantor et al., submitted). The method that originally demonstrated domain-limited repair was used to isolate molecules representing the larger repaired domains, by recovering the DNA from sucrose gradients. These repaired DNA molecules (about 15% of the total) averaged 50 kb in size. This species was found by hybridization to be considerably enriched for β-actin and DHFR sequences, but not for the 754 locus or repetitive sequences.

The different results obtained for repair of the DHFR gene might relate to Cleaver's observation that domain-limited repair is absent from cycling cells. Mullenders et al. (pers. comm.) have observed less efficient repair in the DHFR and ADA genes in XP-C when cells were actively growing. Perhaps repair in the XP-C cell strains used by Bohr et al. is particularly sensitive to the growth state.

Although the possibility remains that the sensitivity of XP-C cells to UV results from a limited capacity to repair vital sequences, further analysis of the relation of such repair to cell survival is warranted. In fact, XP-C cells exhibit considerable resistance to UV when assayed in the arrested state. Their ability to maintain cell metabolism and attachment to the substratum after UV (Kantor and Hull 1984) is much greater than that of XP-A or XP-D cells. Increased survival of XP-C compared to XP-D cells has also been noted by Cleaver and Thomas (1988), using a method that determines incorporation of labeled hypoxanthine in cells incubated for several days after irradiation. XP-C cells also exhibit greater survival when irradiated and held in a quiescent state for many hours prior to plating for assay of colony-forming ability than when plated immediately after irradiation. This response is often termed "liquid-holding recovery" by analogy to experiments in which bacteria are held in buffer to prevent DNA replication after irradiation. XP-A cells do not show such recovery. However, whether XP-C cells demonstrate more proficiency for this process than XP-D cells is questionable (Maher et al. 1979; Chan and Little 1979); careful comparisons using a range of doses and recovery times would be necessary to determine this. Recovery of RNA synthesis after UV is also greater for XP-C than for XP-A, XP-D, or XP-G (Mayne and Lehmann 1982), which is consistent with greater ability to repair active genes. Recovery of DNA synthesis has also been reported to be better for XP-C than for XP-A or XP-D cells (Cleaver et al. 1983; Moustacchi et al. 1979). These recoveries were of course measured in growing cells, and only limited analysis was presented for XP-C cells as the focus of these investigations was on other issues.

Taken together, all these observations demonstrate that repair in XP-C cells is more effective at ameliorating DNA damage than is the case for other repair-deficient human cells. Their much greater UV sensitivity compared to rodent cells in colony-forming assays may be due to several factors, alone or in combination. These include: (1) a lower capacity for repair in vital regions than rodent cells, perhaps manifested to a much greater degree in cycling cells; (2) lower efficiency of tolerance mechanisms, necessary to allow replication of DNA containing persisting damage; and (3) a deficiency in repair of other photoproducts, notably the 6–4 product, as discussed below.

If it is the preferential repair of active DNA in XP-C that brings about its association with the matrix and domain-limited repair, why are these not observed with normal human or rodent cells? For human cells under most experimental conditions, only a small fraction of the repair occurs in active DNA. More recent studies (Harless and Hewitt 1987; Mullenders et al. 1988) indicate that an association of repair synthesis with the nuclear matrix in normal cells does occur at low doses and short times after UV, conditions that might be expected to maximize the proportion of repair occuring in active DNA.

What about rodent cells? Mullenders et al. (1986) observed neither domain-limited nor matrix-associated repair in Syrian hamster embryo primary cells, which exhibit generally poor removal of pyrimidine dimers, like CHO and mouse cells. These cells were chosen because they are particularly suited among the available rodent lines for study in the contact-inhibited state, in which replication is suppressed. A likely explanation for these results relates to the repair of another UV photoproduct, the pyrimidine-(6–4)pyrimidone. Both matrix associated and domain-limited repair are detected by labeling the repair synthesis that accompanies excision repair, and are therefore not limited to repair of PD, while direct examination of repair in specific sequences has for the most part been carried out with TEV, which is specific for this particular lesion. For rodent cells, it is likely that a considerable fraction of the repair synthesis, especially that occurring during short intervals after high UV doses, reflects the repair of 6–4 products. These lesions have been reported to be removed very efficiently and relatively rapidly (Mitchell et al. 1985). Even if initially they comprise only a minor fraction of the lesions, their repair could account for much of the repair synthesis observed. Unless their production were confined to active DNA, which seems unlikely (Lippke et al. 1981), their repair would not appear to be clustered or associated with the matrix. Additional evidence that a large fraction of the repair synthesis observed in rodent cells is unrelated to dimer removal comes from experiments with hamster mutant cell lines. V-Hl cells, derived from V79 cells, exhibit little or no removal of dimers, but remove about half as many 6–4 products as the normal cells (Mitchell et al. 1989). In the first few hours after UV, repair synthesis in the mutant was diminished to about half that in the normal cells (Zdzienicka et al. 1988). Similarly, in the CHO mutant UV 61, which appears to be nearly totally deficient in removing dimers, the amount of repair synthesis relative to the parental cells is not decreased to as large an extent (Van Duin et al. 1988). XP-C cells, on the other hand, appear to be as deficient in removing 6–4 products as dimers (Mitchell et al. 1985), thereby allowing repair synthesis to reflect accurately the amount and genomic distribution of dimer-specific repair.

It appears from these considerations that the mechanisms by which XP-C cells and rodent cells exclude large regions of their genomes from repair activity are not identical. In rodent cells, this exclusion may well apply only to cyclobutane-type dimers, while in XP-C it may apply to several bulky lesions. Unlike the case for rodent cells, the repair heterogeneity in XP-C is influenced by the cell growth state.

Finally, we should note two other interesting observations concerning XP-C cells. Tyrrell and Amaudruz (1987) carried out experiments in which growing cells, seeded at the densities used to measure colony formation, were irradiated and incubated with different concentrations of aphidicolin for 48 h prior to addition of fresh medium for eventual colony assay. For most cells, including normals and XP variants, this treatment led to very large decreases in survival, presumably by blocking repair during the critical early period after irradiation (Keyse and Tyrrell 1987). No sensitization occurred with XP-A cells, consistent with the notion that no repair takes place in them. Two XP-C strains also showed no sensitization, even though they were much more resistant to UV than were the XP-A cells. Another XP-C strain and an XP-D strain showed slight sensitization. Unfortunately, the interpretation of these experiments is made difficult by the fact that the degree of inhibition of repair by aphidicolin varies with the growth state of cells (discussed in detail in Collins et al. 1984) and the effects on the XP-C cells may have been atypical. Similar studies using arrested XP-C, like those performed for other cell types by these authors, would be valuable and should help decide whether an aphidicolin-insensitive pathway for "biologically effective excision repair" as concluded by these authors actually exists in XP-C. Confirmation of this hypothesis would represent an important difference between repair of presumably vital sequences in XP-C and normal cells. Experiments with rodent cells would also be instructive. Another provocative aspect of XP-C is the observation that at least one cell strain is apparently not hypersensitive to the damaging agent 4-nitroquinoline 1-oxide (4NQO), as are the other XP groups, when measured with an assay that only requires limited cell growth (Edwards et al. 1987). However, as the authors point out, other XP-C strains exhibit decreased repair synthesis after 4NQO treatment, and this may be an example of heterogeneity in the complementation group.

4.2 Cockayne's Syndrome

Like XP, Cockayne's syndrome (CS) is a rare disorder with many manifestations, including sun sensitivity. Fibroblasts cultured from patients with CS are sensitive to UV. Although many standard assays failed to reveal any defect in removal of UV damage from the genome overall in these cells, Mayne and Lehmann (1982) obtained evidence for a defect in processing damaged DNA. They showed that in normal cells the rate of recovery of RNA synthesis after UV irradiation is rather rapid, certainly faster than the overall rate of pyrimidine dimer removal, but that CS cells did not exhibit this recovery. They suggested from this that normal cells preferentially remove lesions from active genes and that this process is defective in CS cells.

The confirmation of more rapid removal of pyrimidine dimer from the DHFR gene in human cells (Mellon et al. 1986) provided direct evidence of such preferential repair. Mayne et al. (1988) reported direct observations of removal of TEV-sensitive sites from the ADA gene in normal and CS cells. They showed not only that the rapid repair in the ADA gene (similar to that previously found for DHFR) was absent in the CS cells, but also that the final extent of repair (measured after 24 h) was markedly less in the CS cells. These points are supported by results from preliminary studies of the c-*abl* gene in normal and CS fibroblasts (I. Mellon, unpublished).

Additional insight into the defect in CS comes from analysis of the association of repair synthesis in these cells with the nuclear matrix. In marked contrast to the observations for XP-C cells and normal cells (at short times after low doses of UV), the repair synthesis in CS cells appears to be *depleted* in DNA associated with the matrix (Mayne et al. 1988; Mullenders et al. 1988). Data have not been presented for these kinds of experiments at later times after irradiation.

These observations suggest more than just a defect in *preferential* repair of active genes in CS cells; their repair may be deficient even when compared to the overall genome. This has implications for the mechanism for preferential repair but also indicates that the additional resistance to UV of normal cells compared to CS cells may be the result of more than just the contribution of increased *rate* of repair of active DNA.

Tyrrell and Amaudruz (1987) reported similar features of aphidicolin treatment on survival of UV-irradiated CS and normal cells. In contact with inhibited cells, a 48-h incubation in the drug markedly sensitized both types, but it was only the first few hours after UV in which the drug was effective. At first, this might seem to be inconsistent with the notion that CS cells are defective in a rapid-repair process that confers UV resistance on normal cells. However, in order to do the experiment, the CS cells were irradiated with a very low dose. If a large fraction of killing by UV is due to persistence of transcription-blocking lesions in some genes coding for proteins critical for metabolism and which turn over rapidly, inhibiting the repair that does occur in CS cells would still result in increased killing.

4.3 Repair in Complemented Mutants

Removal of pyrimidine dimers from the DHFR gene has also been examined in CHO repair-deficient mutant cells and transformants of these cells that carry heterologous repair genes. Two different mutants have been examined. Cells of mutant line 43-3B (complementation group I), originally derived from the line designated CHO9, exhibit greatly increased resistance to killing by UV when carrying the human gene, ERCC-1. Cells of the mutant line UV-5 (complementation group II), derived from the AA8 line of CHO, show a lesser, but still significant increase in resistance to UV when expressing the phage T4 gene encoding TEV. Cells complemented by the human gene exhibit the preferential repair characteristic of the wild-type CHO. Bohr et al. (1988) reported that the ERCC-1 gene expression restored repair in the DHFR gene to wild-type levels, but the repair in the downstream sequence remained low. Mellon et al. (1988) found that the strand specificity of repair was restored in the mutant carrying ERCC-1. Thus it appears that the product of the human gene functions in the CHO repair system in the same way as the deficient CHO product.

In contrast, TEV does not restore preferential repair. Bohr and Hanawalt (1987) compared the repair in the 5' half of the DHFR gene to that in the sequence immediately downstream from the gene in AA8 cells and their derivatives in the 8 h following 20J/m^2 UV. As expected, the wild-type cells exhibited efficient repair in the gene but not in the downstream sequence, but the mutant cells showed little repair in either. In the cells expressing TEV, *both* sequences were repaired to a slightly greater extent than was the gene in the wild-type cells. Hanawalt et al. (1989) reported examination of strand specificity of repair in the 5' fragment in the gene in the same cells after 10 J/m^2. They observed strand specificity in the wild-type cells, although repair in the transcribed strand was only 50% after 8 h, somewhat less than reported for other lines. The cells expressing TEV repaired the transcribed strand to about the same extent, but they also repaired the non-transcribed strand. TEV is a pyrimidine dimer glycosylase; its substrate is the pyrimidine dimer itself, not DNA distortion. Perhaps TEV is capable of recognizing PD in the absence of any additional signals that are required for recognition by the endogenous nucleotide excision repair system. A similar lack of preferential repair or strand specificity has been found for repair of N-methylpurines, whose repair is also mediated by a small glycosylase (see below).

5 Repair of Other Lesions

To this point we have only discussed repair of the UV-induced cyclobutane pyrimidine dimer. General conclusions about DNA repair have often been made from studies using UV as the damaging agent, and the present case provides no exception. However, it is critical to a thorough understanding of the biological role and mechanism of heterogeneous repair to determine the extent to which repair of other lesions resembles that of PD. This can provide insights into which aspects of lesions may be important for preferential repair, e.g., the ability to block transcription or distort the DNA, and whether other repair systems exhibit selectivity for certain regions of the genome. Although there is compelling evidence that many aspects of the repair pathways for different bulky adducts are similar, it is becoming increasingly clear that they need not be identical. Human and rodent repair-deficient mutants classified into the same complementation group, based on responses to UV-irradiation, may exhibit different degrees of sensitivity to chemical damaging agents or deficiencies in removing a given lesion. Mutants are being found that appear to be deficient in removal of one particular bulky lesion but not another (Thompson et al. 1989; Cleaver et al. 1987).

In fact, rodent cells in culture, which might be formally described as "mutants" if their generally inefficient repair of PD is a culture artifact (Peleg et al. 1977), may not exhibit this deficiency for other adducts. The rapid and efficient removal of 6–4 photoproducts that has been reported (Mitchell et al. 1985) for rodent (and human) cells in culture suggests a possible resolution of the rodent paradox, different from the one entertained in the beginning of this chapter. In this proposal, the similar UV-resistance of human and rodent cells stems from similar rates and extents of repair of these 6–4 photoproducts, which, like repair of PD, depends on a functional excision repair system. The subject of the relative importance of PD and 6–4 photoproducts for survival and mutagenesis remains controversial. The extreme positions that either product is solely responsible for the adverse effects of UV seem to have little likelihood of being correct. However, the evidence that the 6–4 product is rapidly and efficiently repaired in all sequences in rodent cells has important implications for the nature of the "defect" in these cells in PD removal. Study of the removal of 6–4 products is confounded by the fact that they are a minor product, no agent is known that produces them preferentially, and no specific endonuclease for them is currently available. Most data about their removal is based upon a competitive radioimmunoassay that uses a radioactive ligand having very high specific activity and a polyclonal antibody. It would be reassuring to corroborate the essential features of their repair by an independent method. Perhaps by studying other bulky adducts, we can place heterogeneity of repair of PD in proper perspective.

In comparing repair in different sequences in this case, care must be taken to establish whether the sequences differ in susceptibility to adduct formation. This is especially important for comparisons of active and inactive DNA, since numerous examples of increased binding of chemicals to active chromatin or ribosomal DNA have been reported (see Bohr et al. 1987).

The identification of preferential repair for PD differed for human and rodent cells; there is a good consensus among investigators that, at biologically relevant doses, the former remove most of these lesions from their genomes and the latter remove only a small fraction within 1–2 days. These characteristics served to define a baseline against which repair in a specific sequence could be compared. Expectations of what constitutes "preferential" repair of a sequences for a chemical adduct must also be based on its overall repair characteristics. Unfortunately, such a consensus does not exist for bulky chemical adducts, a class defined not only by the chemistry of the products but also by the spectrum of damaging agents to which well-characterized DNA repair mutants are sensitive. A survey of the literature results in a confusing array of sometimes contradictory results concerning the rate and extent of removal of bulky chemical adducts from both human and rodent cells. Often, comparisons that seem desirable now must be made using data obtained by different investigators, who usually approached the measurements with different objectives. There is also a tendency to regard recent investigations more favorably than prior ones in the case of discrepancy, but this must be carefully justified. All we can do here is present the little available data about removal of chemical adducts from specific sequences and discuss it in light of previous studies of the overall genome.

5.1 Rodent Cells

5.1.1 Acetylaminofluorence Adducts

The most direct comparison available for preferential repair of UV damage and chemical damage is a study carried out with the amplified DHFR locus in the CHO cells originally used by Bohr et al. (1985). In this study (Tang et al., 1989), the UVR-ABC nuclease of *E. coli* was used to incise DNA at the sites of lesions formed by N-acetoxyacetylaminofluorene (N-AAAF). HPLC analysis of labeled adducts in the genome overall showed that only the C-8-AAF adduct to guanine was formed in appreciable quantities. Control experiments showed near-quantitative incision at these adducts by the UVR-ABC nuclease, and indicated the same linear dose response for adduct formation in the 14-kb fragment in the DHFR gene and in the 14-kb fragment downstream from the transcription unit. A dose of 10 uM N-AAAF resulted in about three enzyme-sensitive sites per 14 kb of DNA in the two specific sequences and 2.5-3 adducts per 14 kb in the overall genome, as measured by HPLC. Removal of adducts from the genome overall and from the two specific sequences followed the same kinetics, with 60-70% removed in 24 h, and about 40% removed after 6 h, the earliest point included.

These results are clearly dissimilar to those reported for removal of pyrimidine dimers in these cells. Removal of AAF adducts appears relatively efficient over the whole genome, and at the times examined, repair in the DHFR locus resembled overall repair. This situation more closely resembles the removal of pyrimidine dimers by human cells. The key question of whether repair is more rapid in the gene, especially in the transcribed strand, awaits further study. It is perhaps worth noting that the initial lesion frequency used was considerably higher than in most experiments with UV; this might lessen the apparent differences in rate of repair between transcribed and silent sequences.

In previous studies, Poirier et al. (1979) reported that confluent mouse and human epidermal cells treated with N-AAAF removed about 50% of the adducts (mostly C-8-AF adducts) in 24 h, using a radioimmunoassay, whereas Amacher et al. (1977) reported significantly less removal of labeled adducts from mouse fibroblasts than from human fibroblasts.

5.1.2 Psoralen Adducts

The formation and removal of psoralen DNA interstrand crosslinks and crosslinkable monoadducts in specific sequences has been studied by exploiting the fact that crosslinked molecules can be selectively returned to duplex form after denaturing treatments, allowing their separation from the single-stranded, uncrosslinked molecules. Restriction fragments can be separated by electrophoresis, transferred to support membranes and analyzed by hybridization (Vos 1988). In the case of psoralen, crosslinks are produced in a two-step photoreaction, whereby intercalated molecules absorb one quantum to produce monoadducts, and then a considerable fraction of these monoadducts (65%) are able to absorb another quantum to form crosslinks. The frequency of crosslinks can be measured directly; the frequency of the crosslinkable monoadducts can be determined from the additional crosslinks produced by irradiating the purified DNA prior to analysis.

Most of the work using this method has been done with human cells and is discussed below. Unpublished studies with CHO cells (Vos et al., in preparation) indicated that the removal of the monoadducts of the soluble psoralen derivative hydroxymethyl-trimethylpsoralen from the DHFR gene appears analogous to the removal of pyrimidine dimers. About one-half the crosslinkable monoadducts were removed in 8 h. Since monoadducts on either strand can produce crosslinks, this could result from rapid removal of adducts on the transcribed strand. Unfortunately, data have not been obtained for the unexpressed downstream fragment to allow study of the possible involvement of transcription in repair of these adducts. Little information is available about removal of these adducts from the genome as a whole. In the only report available for CHO cells, Ben-Hur and Elkind (1973) measured the disappearance of labeled trimethylpsoralen from extracted DNA, and concluded that psoralen adducts were rapidly and efficiently removed (80% in 8 h). Szafarz et al. (1983) reported that 40% of labeled 3-carbethoxypsoralen monoadducts were removed from the DNA of mouse embryo fibroblasts in only 1 h. Additional study is clearly needed to compare repair in the overall genome and in specific sequences using the same agent and adduct frequencies.

Removal of DNA crosslinking from the DHFR gene was found to be more rapid and extensive than that of monoadducts, being nearly complete at the first time examined, 8 h. For ribosomal genes however, the results were considerably different. A slow removal of crosslinking was observed, reaching only about 50% in 48 h, and monoadduct removal was even slower and less extensive.

5.1.3 Other Bulky Adducts

Previous studies with some other carcinogens also suggest that rodent cells in culture are not markedly deficient in removing bulky adducts, other than PD, when compared to human cells. Thompson et al. (1984) examined the removal of 7-bromomethylbenz[a]anthracene adducts in CHO AA8 cells and in a number of repair-deficient mutants, using labeled carcinogens. They found 50–60% removal in 24 h in wild-type cells, but less than 15% removal for mutant cells from five complementation groups that show sensitivity to the carcinogen (and UV as well). Dipple and Rogers (1977) compared repair of these adducts in human and Chinese hamster cells, and found that at low doses 40–50% of total adducts were removed from the cells in 30 h. Ikenaga and Kakunaga (1977) compared the removal in mouse and human cells of various adducts produced by treatment with 4-nitroquinoline 1-oxide. The stable purine adducts, which are not removed by human XP-A cells, were removed to the same extent (50% in 24 h) in both mouse fibroblasts and human amnion cells.

5.1.4 Methylated Purines

Many carcinogens and mutagens do not form bulky adducts to DNA but make chemical changes in the structure of the bases themselves. Such damage is removed by specific glycosylases each of which recognize and remove one or a few types of altered bases, leaving abasic sites. These sites are then subject to excision repair initiated by AP endonucleases. This appears to be the case for 3-methyladenine and 7-methylguanine, two major products of simple methylating agents. Even though 7-methylguanine itself may be relatively innocuous, its spontaneous release from DNA to create mutagenic AP sites may provide a reason for its active removal. The heat lability of these two products in DNA has been exploited to measure their formation and removal in the DHFR locus of CHO cells treated with dimethyl sulfate (Scicchitano and Hanawalt 1989; Hanawalt et al. 1989). Rather than treat the isolated DNA with an endonuclease, they heated it to produce AP sites and then incubated it in alkali to cleave DNA at these sites. The remainder of the procedure was the same as that developed by Bohr et al. (1985) except that to provide an "untreated control" portions of the samples were heated in the presence of methoxyamine, which reduces the AP sites formed, making them refractory to alkali. Thus, any effects of extraneous alkali-labile sites on the analysis were eliminated.

The initial formation of DMS-dependent, heat-labile sites was found to be the same in the DHFR gene and the silent downstream sequence. Repair was readily observed, reaching 70–80% after 12 h and virtually complete at 24 h. This result was obtained both for the gene and its flanking sequence, as well as for the two individual strands of the gene itself. At 3 and 6 h, a small increase in the repair in the transcribed strand relative to the opposite strand was observed, but this was not deemed significant.

Repair in the overall genome was measured by determining molecular weights of similarly treated DNA using alkaline sucrose gradients. The initial frequency of lesions was about 25% lower than observed in the DHFR locus, and repair appeared slightly slower and less complete, reaching only 72% in 24 h.

These results may be interpreted in terms of the repair system used and the effects of the lesions on DNA transactions. Chromatin structure may not restrict access of DNA to the small glycosylases and AP endonucleases involved in this repair system. In addition, if the blockage of transcription by pyrimidine dimers plays a role in their recognition in CHO cells, the lack of such a blockage by the major adduct studied here, 7-methylguanine, could help account for the lack of difference between transcription template and nontemplate strands.

It should be noted that some evidence exists for intragenomic heterogeneity in removal of 0^6-metylguanine, another product of small methylating agents, in rat liver. Ryan et al. (1986) observed both greater production and removal of this product in a chromatin fraction of rat liver enriched for active sequences than in the bulk of the genome. A "nuclear matrix" fraction was very poorly repaired; the relation of this fraction to the nuclear matrix defined by the methods used by Mullenders et al. is unclear. Removal of N-methyl purines was about the same in all the fractions. The 0^6-methylguanine is repaired by direct reversal via an alkyltransferase protein; that activity might be concentrated in active chromatin regions.

5.2 Human Cells

5.2.1 Aflatoxin

Repair in metallothinoein genes in human cells treated with activated aflatoxin B_1 was studied by Leadon and Snowden (1988), using antibody to BrUra in DNA. As was the case for UV, at 6 h after damaging treatment, the fraction of restriction fragments in the antibody-bound fraction that contained active genes was a little more than twice that found for the bulk of the DNA, while at later times the differences were not very significant. Unlike the case for UV, the fragments containing unexpressed genes appeared to be repaired less well than the bulk of the DNA.

On the surface, these results suggest that preferential repair of AFB_1 adducts occurs in the active MT genes in human cells. The significant question remains, however, of the degree to which the initial distribution of adducts may bias these results. AFB_1 adducts have been shown to be more prevalent in the nucleosome linker DNA (Kaneko and Cerutti 1980) and in active ribosomal DNA (Irwin and Wogan 1984). A number of chemical adducts have been shown to occur in higher frequency in chromatin fractions thought to be enriched in transcriptionally active sequences than in bulk DNA (reviewed in Bohr et al. 1987), although such studies apparently have not been reported for aflatoxin. In addition, formation of these adducts in the nontranscribed alpha DNA of green monkey cells in the nondividing state was not different from that in bulk DNA, although their enzymatic removal from alpha DNA was deficient (Leadon et al. 1983).

5.2.2 Psoralen Adducts

Vos and Hanawalt (1987) reported studies of the removal of psoralen monoadducts and crosslinks from the DHFR gene of the same cell line, containing multiple copies of the gene, used by Mellon et al. (1986; 1987) to study pyrimidine dimers. By 8 h after treatment, only about 20% of the crosslinkable monoadducts had been removed from a large (32-kb) restriction fragment encompassing the entire gene. Removal was about 50% complete in 24 h. No measurements were made of removal in the genome overall or in a silent sequence in this study; to assess whether preferential repair occurred we must use data obtained in other studies. The lack of complete removal of psoralen adducts has been found generally (Smith 1988b; Vuksanovic and Cleaver 1987), and 50% removal at 24 h is not an atypical result. When compared to the removal of pyrimidine dimers, 20% removal at 8 h appears too low to signify preferential repair. The method used is capable of detecting adducts at much lower frequencies than the endonuclease method (in the range of 0.1 per fragment) and the values for removal at 24 h were similar at both high and low frequencies. However, data at 8 h were only reported for relatively high adduct frequencies (0.5 crosslinks, 1.55 crosslinkable monoadducts, and probably about 0.8 noncrosslinkable monoadducts per duplex restriction fragment). At these frequencies, a significant number of genes contain both monoadducts and crosslinks. This may obscure detection of preferential repair of the monoadducts. However, a similar value for removal was recently also obtained using conditions that produce fewer adducts (Vos et al., in preparation). In addition, psoralen monoadducts in the nontemplate DNA strand are probably not blocks to transcription (Shi et al. 1987), and since the overall rate of removal of these adducts appears to be significantly less than that for removal of pyrimidine dimers, their removal may be very limited at short times. Since repair of these lesions was not (and cannot be) measured on the individual strands with the method used here, this may have also obscured detection of preferential repair of adducts on the template strand.

Although removal of crosslinking from the gene was more extensive (75–80% in 24 h), it too was only around 25% in 8 h. One would expect crosslinks to be effective blocks to transcription. This is indeed the case for T7 (Shi et al. 1988) and *E. coli* (Shi et al. 1987) RNA polymerase in vitro. Unless the presence of multiple adducts in the fragments affects the processing of crosslinks, this suggests that incision at sites where transcription has been blocked by a crosslink is less efficient than at those sites blocked by dimers. In the assay used here, all that is necessary for the crosslinking to be lost is incision (or some other means of breaking the bridge); full restoration of DNA integrity is not required. Perhaps the structure at the site of transcription blockage is fundamentally different at crosslinks and dimers, and the former do not provide the proper configuration for incision of the DNA. Psoralen adducts might also cause some general inhibition of transcription, not mediated by direct blockage of the polymerase, lowering the frequency at which lesions might be made more available for repair. Viewed from another standpoint, the adduct frequency in the gene itself is likely to be considerably greater than in nontranscribed sequences (Ross and Yu 1988; A. Islas, unpublished observations) so the amount of processing in the gene at early times may in fact be greater than that occurring in silent sequences. Islas et al. (in preparation) have separated crosslinked and uncrosslinked DNA molecules, according to their buoyant densities after denaturing-renaturing treatments, and probed the two classes for content of various sequences. They observed more efficient removal of crosslinking from the DHFR gene than from the bulk DNA, and considerably less efficient removal from an inactive proto-oncogene, c-*fms*.

For human cells, removal of psoralen adducts in the ribosomal genes, measured by Southern analysis, was even less than that in CHO cells, being undetectable for monoadducts and barely detectable for crosslinks. Interestingly, removal of crosslinking mediated by mitomycin C treatment of human lymphoblastoid cells was not markedly deficient, except in some lines from Fanconi's anemia patients (Matsumoto et al. 1989).

The technique of converting monoadducts to crosslinks in purified DNA was also used to measure the presence of these lesions in replicated DNA. When replicated DNA isolated from the density gradients was re-irradiated and analyzed, the DHFR genes were found to contain crosslinkable sites at a frequency great enough to argue that replicational bypass of these lesions must be very efficient (Vos and Hanawalt 1987). Subsequent experiments have shown that such bypass also takes place in the ribosomal genes of both human and CHO cells, sequences which seem to be poorly repaired. This raises the possibility that replication of genes containing damage provides an alternative means for restoring transcriptional competence to vital genes. Although prone to error, the chance that the template generated in this way would still produce a functional protein would be relatively high, due to the large amount of noncoding sequence that makes up most mammalian genes, and the fact that many possible sequence changes would result in no (or benign) amino acid substitutions. Whether this would also be true for ribosome function is less clear. Unless induced by damage, this process would only occur in dividing cells. It should be kept in mind that the efficient bypass of monoadducts during replication has so far been demonstrated only in situations in which tandem repeats of the sequence occur. If recombinational processes are involved in such bypass, its efficiency may be abnormally high in the cases studied so far.

5.2.3 Methylated Purines

Nose and Nikaido (1984) examined production and removal of single-strand breaks and alkaline-labile sites in restriction fragments hybridizing to probes for the collagen I gene and β-globin gene in confluent human fibroblasts. The fragments were 4 and 5.2 kb, respectively. After treatment with increasing doses of methylnitrosoguanidine, the amount of intact fragments declined to about 30% at the highest dose used (20 μg/ml for 1 h), corresponding to about 1.2 sites per fragment. To examine repair, they used a much smaller dose (1 μg/ml), which resulted in about 0.35 sites per fragment in DNA isolated immediately after treatment. However, in DNA isolated 4 h later, the number of sites had increased to about 0.9 per fragment. At about 18 h, the frequency of sites was about 0.1, and at 30 h no sites were found. These results were similar for both the collagen gene, for which they demonstrated the presence of homologous polyA containing RNA in the cytoplasm, and for the globin gene, for which no RNA was found.

The similarity of results for the two different genes are in agreement with those of Scicchitano and Hanawalt (1989) for CHO cells, assuming that the lesions under study were in fact AP sites formed from N-methylpurines in the DNA. The increase in sites at 4 h suggests cellular processing of these lesions to form relatively long-lived AP sites. It is not clear whether this study would have been able to reveal differences in the rate of processing of lesions at early times after their formation. The overall kinetics of repair of the alkaline-labile sites were not investigated by these authors.

6 Mutagenesis

Intragenomic heterogeneity in repair processes could influence mutagenesis to a variety of degrees, depending primarily upon the extent of deficiency of repair that can occur. In the case of pyrimidine dimers in growing rodent cells in culture, the apparent lack of repair in unexpressed genes would be expected to result in higher mutation frequencies in them, compared to housekeeping genes. If cells in the organism exhibit the same sort of repair heterogeneity for a biologically relevant mutagen, the lack of repair in genes whose expression is inappropriate for the function of the exposed cell could have numerous consequences. Mutations in some genes could be exposed in the course of differentiation, affecting proper development or tissue-specific function. Depending upon the precise mechanisms later used to suppress continued cell division, such mutations in unexpressed genes could later result in carcinogenesis. DNA rearrangements fostered by unrepaired damage could have similar results. The unrepaired damage in this case need not be in an inactive gene; it could be in a silent region of the genome.

For obvious reasons, expressed genes have been used for analysis of the nature and frequency of mutations. To compare mutation frequency in expressed and silent genes would require development of a specific test system. To adapt usual methods, one would need an inducible gene whose loss of activity could then be used as the basis for selection under defined conditions, which would have to be applied after mutagen treatment and a suitable interval to allow for mutation fixation. Even if such a system were available, mutations that interfered indirectly with the system might be a serious impediment.

We might expect the lack of repair in the nontemplate strand of a gene, demonstrated thus far only for DHFR, to be reflected in a strand bias in mutations detected in it. However, it must be kept in mind that if the means for identifying mutants depends upon loss of the functional gene product, the sites at which a sequence change caused by a specific mutagen would lead to such a result may also be biased with respect to the strands, and could influence the result. To identify strand bias, the exact nature of mutagenesis brought about by the mutagen used must be known to allow the deduction of the location of the lesion that led to the mutation.

For situations where repair heterogeneity only involves rates, the degree to which it will be reflected in mutation depends directly on the time available to the cell prior to mutation fixation.

For practical reasons, detailed analysis of mutations by treating mammalian cells with UV or chemical mutagens has been confined to genes whose small size facilitates analysis of sequence changes. Larger genes have been chosen for analysis of repair, also for technical reasons. Recently, however, mutation analysis in a large gene has been made possible by analyzing the RNA produced by mutant cells, rather than the DNA itself (Vrieling et al. 1988). When this method was applied to the Chinese hamster HPRT gene, some very interesting results were obtained (Vrieling et al. 1989).

In the V79 cells analyzed, nearly all of the UV-induced mutations in HPRT recovered were base changes located at positions consistent with mutations caused by replication errors at di-pyrimidine photolesions. Assuming this to be the mechanism for mutation, the strand containing the lesion could be identified. In repair proficient cells, 9 of 15 mutants appeared to result from lesions in the nontranscribed strand. This slight bias is probably not statistically significant. In the mutant V-H1, however, an extreme strand bias in favor of the transcribed strand was observed; 7 of 8 mutations appeared to result from lesions in this strand.

The V-H1 cells have been characterized as extremely deficient in removing pyrimidine dimers both in the genome overall and in the HPRT gene itself, with some residual repair of 6–4 photoproducts. It would therefore be expected that little or no strand bias in the frequency of lesions encountered by the replication apparatus would occur. A plausible explanation for the strong bias in mutant frequency observed is that mutation fixation by replication itself exhibits a strand bias, being more prone to inserting incorrect bases when copying the transcribed strand. This could be due to different polymerases or constraints involved in replicating the leading versus the lagging strand. Assuming that these differences are unrelated to excision repair, and therefore should also operate in repair-proficient cells, one can argue that the mutational bias toward the other strand observed in these cells indicates very efficient repair in the transcribed strand of the HPRT gene. For this gene then, repair confined to the transcribed strand has the effect of lowering the mutation rate far beyond the factor of two that might be predicted in the absence of different frequencies of mutagenic sites on the two strands. If the fixation bias relates to continuous and discontinuous replication, some genes would likely be in the other orientation, such that repair would have little or no effect on mutation frequency. One might predict that for some genes, mutation frequency would be the same in V79 and the V-H1 mutant. There is the possibility that genes are not randomly distributed with respect to this polarity. In *E. coli*, genes tend to be arranged with respect to the single origin of replication in such a way as to replicate the transcribed strand in the continuous mode (Brewer 1988). Recently, it has been shown in yeast that the ribosomal DNA, although it contains many origins, is functionally replicated in only one direction; replication forks appear to terminate before they proceed into the genes in a direction opposite to transcription (Brewer and Fangman 1988; Linskens and Huberman 1988). Whether this occurs for other eukaryotic genes is unknown.

These results underscore the hazards of trying to correlate mutation and repair. Had the mutant cells not been analyzed, this study would have provided only very weak evidence for strand-specific repair. Thus at present, data concerning other genes, cell types, and mutagens cannot readily be used to make inferences about the repair process.

7 Conclusions and Perspectives

Even before methods were developed to examine active genes, comparisons between the nontranscribed, repetitive alpha DNA of monkey cells and the remainder of the genome (reviewed by Smith 1987) had already demonstrated how complex repair heterogeneity could be. Removal of psoralen and aflatoxin adducts was markedly dedicient in alpha DNA, while removal of pyrimidine dimers, thymine glycols, and methylated purines was not. Furthermore, the repair heterogeneity was characteristic only of nondividing cells. These results were interpreted to mean that the chromatin structure of this centromeric, heterochromatic DNA in nondividing cells severely restricted access of the repair system for bulky chemical adducts. UV radiation, however, somehow relaxed this restriction, promoting removal of both chemical and UV-induced lesions, suggesting that it stimulated some active process to "open up" alpha chromatin for repair. The apparent unrestricted access of glycosylases was attributed to the small size of these proteins.

That chromatin should provide barriers of varying degree to enzymes whose substrate is the DNA itself is nearly axiomatic. Since alpha DNA is not transcribed, the postulated changes in its chromatin might represent an extreme case of relaxing a very compact structure to that characteristic of the bulk of the DNA. The existence of "preincision" activities necessary for repair of bulky damage over the entire genome has been postulated many times, reconciling a variety of observations. Since chromatin exists at several different degrees of compaction, variable amounts of processing are probably required to bring different domains to a level corresponding to the substrate for the components of the repair system that actually recognizes DNA damage. It seems reasonable to assume that repair would require the same degree of chromatin decondensation as transcription or replication, and that this would be at the nucleosome level. In this context, it is easy to envision transcriptionally active DNA to be in a chromatin configuration more readily repaired than highly condensed DNA like alpha in nondividing cells.

The preperential repair of the DHFR gene in CHO cells seemed easily interpretable in these terms. If most of the DNA in the genome exists in domains that require opening up for repair, a deficiency in this function could lead to a restriction of repair to DNA already in an opened configuration, such as the DNA of actively transcribing genes. It was not clear whether entire structural domains were accessible, or whether further chromatin modification characteristic only of transcription units themselves was needed for repair. With this model, the more rapid repair of active genes in human cells could be seen as a consequence of the extra time required for decondensing inactive chromatin, a step proficient in these cells. The domain limited repair in XP-C cells suggested that they, like rodent cells, were defective in some aspect of this preincision processing function. Another explanation was that a limitation in total repair capacity in XP-C results in the repair of only certain, more accessible domains.

The strand specificity of repair in the DHFR genes of CHO cells challenges this general accessibility model and forces consideration of repair at the nucleosome level, where chromatin could still pose a barrier to recognition of damage. The recent elucidation of many details of the mechanism of action of the *E. coli* UVR-ABC recognition-incision complex on damaged DNA in vitro has led to a specific model for its action in vivo (Grossman et al. 1988; Sancar and Sancar 1988). Some repair proteins bind to and translocate along the DNA, recognizing certain changes in the spatial path of the helix that result from the presence of a variety of distorting lesions. Other proteins then recognize the unique DNA-protein substrate formed when the translocation ceases at the site of a lesion, and complete the formation of the complex which incises the DNA. A freely diffusible, damage-specific endonuclease, similar to the glycosylases, is simply not part of the strategy used. If eukaryotes use a similar mechanism, the proteins that recognize DNA distortion may need to scan DNA unbound by chromatin proteins. Thus, as well as decondensing structural domains, the repair system might also have to bring about an ordered disassembly and reassembly of nucleosome structure, perhaps using elements analogous to those involved in replication and transcription. A defect at this level could limit recognition of damage, except at sites where chromatin is stably disrupted, such as stalled transcription complexes.

A strict correlation of repair with transcription blockage in rodent cells would rule out repair in nontranscribed sequences. This may in fact be the case for repidly growing cells like CHO; more extensive data on repair in sequences carefully analyzed with respect to transcription will be necessary to decide this issue. Different rules may apply to differentiated cells, or only to rat cells.

For human cells, "preferential repair" is more difficult to determine (and to define) as it depends on rates, which may vary for different reasons among different sequences. The current evidence from studies of the β-actin gene locus suggests that rapid repair may not be strictly confined to transcription units, and it appears that repair in XP-C cells may be confined to structural domains, probably containing active genes, rather than only to the transcription units themselves. Information about strand specificity of repair in these cases is obviously needed. There may be several hierarchies of repair efficiency within human cells: perhaps some domains are so particularly well situated for scanning by the repair system and they are repaired so rapidly and completely that transcriptional activity within them is irrelevant to their repair. In other domains, scanning may be less efficient so that dimers which block transcription are repaired more rapidly than others. Completely inactive or highly condensed domains may be repaired very slowly or not at all.

At present we have only rudimentary data concerning removal of bulky chemical adducts from specific sequences, but it does appear that models to explain repair heterogeneity will require still further refinement. How can 6–4 photoproducts and AF adducts, but not dimers, be recognized and removed throughout the rodent genome? A possibility consistent with the general accessibility model is that local disruption of chromatin structure by these adducts is in itself sufficient to expose the DNA to repair proteins, eliminating the need for opening up the chromatin to remove them. If, in fact, the cells are proficient in opening up and scanning the entire genome for damage, then they must be deficient specifically in removing dimers, except those that block transcription. We can discern two ways this could come about, using the model for UVR-ABC action, in which *recognition* is the proper assembly of proteins at a lesion site, leading to incision, brought about through participation of another protein. One possibility is that the helical distortion at a PD is not sufficient to make it a good substrate for the recognition system, and cells use an additional factor, sensitive directly to the structure of the lesion, to enhance the probability of this event. A deficiency in this factor in rodent cells could then result in inefficient recognition of dimers in inactive DNA, while in transcribed regions, transcription blockage could substitute for the missing function. For example, Patterson et al. (1987) have suggested that cleavage of the phosphodiester bond connecting the bases of a pyrimidine dimer may facilitate or be required for its recognition by human cell repair systems. This is based on the inferred presence of such components both in the genomic DNA of excision-deficient cells and in apparent excision products in normal cells. Until mutants known to lack an enzyme that can perform this cleavage are characterized, one cannot determine whether such processing is a normal part of the repair system or is adventitious. [Such processing alone is not responsible for the loss of PD inferred from the loss TEV-sensitive sites in rodent cells; proficient repair in the DHFR gene of CHO cells has also been demonstrated using the UVR-ABC nuclease (Thomas et al. 1988) which should recognize PD modified in this way. In addition, such cleavage products are not found in quantity in Syrian hamster cells (Pirsel et al. 1989) at early times after irradiation.] Another possibility for such a dimer-specific, ancillary component of the repair system might be a protein that binds to PD at the site of the dimer and enhances recognition, much as the binding of photolyase in the dark enhances UVR-ABC nuclease activity (Sancar and Sancar 1988). If PD are in fact properly recognized, a second possible explanation for a dimer-specific deficiency would involve the subsequent incision step. A defect in one of the components of the repair system might inhibit incision only at the configuration of PD in a preincision complex, but not at a complex formed at the sites of transcription blockage or at other types of lesions. The fact that two hamster mutants that exhibit a greater deficiency in repair of dimers than of 6–4 photoproducts occupy different complementation groups attests to the complexity of the incision process, even within the proficiently repaired domains in these cells.

With human cells, we seem to be confronted with the opposite problem; why are some of these same chemical adducts more poorly repaired than PD? At present, we have even less information about repair of chemical damage in active genes than we have for rodent cells.

A important question is whether preferential repair is an active, directed process, or whether it merely reflects the manner in which repair systems are integrated into the entire gamut of regulated nucleic acid-protein interactions in the cell. The fact that Cockayne's syndrome cells seem to repair active genes more poorly than the overall genome suggests the possibility that there are components of the repair system for active genes that are unnecessary for repair of the rest of the genome. A more detailed understanding of the defect in CS cells will be needed to clarify this issue.

Clearly, much remains to be done. Simple rules governing repair efficiency in specific sequences have not emerged from results of studies with the various model systems used to date. We can hope that with the accumulation of more extensive data concerning repair of different kinds of lesions in a variety of genes, including information about strand specificity of repair, we will be able to distinguish which features of repair heterogeneity reflect general aspects of the organization of repair systems.

Acknowledgements. We are grateful to P. C. Hanawalt for encouragement and advice, and to all the members of the laboratory for helpful discussions. We thank the many investigators who have allowed us to cite unpublished work. Many of the ideas presented here, especially concerning XP-C cells, were developed in long and detailed discussions with L. Mullenders and A. A. van Zeeland, carried out through the facilities of Bitnet/Earn. This work was supported by an Outstanding Investigator Award (CA 44339) from the National Cancer Institute, US Public Health Service, to P. C. Hanawalt.

References

Amacher DE, Elliott JA, Lieberman MW (1977) Differences in removal of acetylaminofluorene and pyrimidine dimers from the DNA of cultured mammalian cells. Proc Natl Acad Sci USA 74: 1553–1557

Ben-Hur E, Elkind MM (1973) DNA crosslinking in Chinese hamster cells exposed to near ultraviolet light in the presence of 4,5',8-trimethylpsoralen. Biochim Biophys Acta 331: 181–192

Bohr VA (1987) Differential DNA repair within the genome. Cancer Rev 7: 28–55

Bohr VA, Hanawalt PC (1986) Novobiocin does not inhibit DNA repair in an active gene. Carcinogenesis (Lond) 7: 1917–1920

Bohr VA, Hanawalt PC (1987) Enhanced repair of pyrimidine dimers in coding and non-coding genomic sequences in CHO cells expressing a prokaryotic DNA repair gene. Carcinogenesis (Lond) 8: 1333–1336

Bohr VA, Hanawalt PC (1988) DNA repair in genes. Pharmacot therapie 38: 305–319

Bohr VA, Okumoto DS (1988) Analysis of pyrimidine dimers in defined genes. In: Friedberg EC, Hanawalt PC (eds) DNA repair. A Laboratory Manual of Research Procedures. Marcel Dekker, Inc, NY, pp 347–366

Bohr VA, Wassermann K (1988) DNA repair at the level of the gene. Trends Biochem Sci 13: 429–433

Bohr VA, Smith CA, Okumoto DS, Hanawalt PC (1985) DNA repair in an active gene: removal of pyrimidine dimers from the DHFR gene of CHO cells is much more efficient than in the genome overall. Cell 40: 359–369

Bohr VA, Okumoto DS, Hanawalt PC (1986a) Survival of UV-irradiated mammalian cells correlates with efficient DNA repair in an essential gene. Proc Natl Acad Sci USA 83: 3830–3833

Bohr VA, Okumoto DS, Ho L, Hanawalt PC (1986b) Characterization of a DNA repair domain containing the dihydrofolate reductase gene in Chinese hamster ovary cells. J Biol Chem 261: 16666–16672

Bohr VA, Phillips DH, Hanawalt PC (1987) Heterogeneous DNA damage and repair in the mammalian genome. Cancer Res 47: 6426–6436

Bohr VA, Chu EHY, van Duin M, Hanawalt PC, Okumoto DS (1988) Human repair gene restores normal pattern of preferential repair in repair defective CHO cells. Nucleic Acids Res 16: 7397–7403

Brewer BJ (1988) When polymerases collide: replication and transcriptional organization of the *E. coli* chromosome. Cell 53: 679–686

Brewer BJ, Fangman WL (1988) A replication fork barrier at the 3' end of yeast ribosomal RNA genes. Cell 55: 637–643

Chan GL, Little JB (1979) Resistance of plateau-phase human normal and xeroderma pigmentosum fibroblasts to the cytotoxic effect of ultraviolet light. Mutat Res 63: 401–422

Cleaver JE (1986) DNA repair in human xeroderma pigmentosum group C cells involves a different distribution of damaged sites in confluent and growing cells. Nucleic Acids Res 14: 8155–8165

Cleaver JE (1987) Relative importance of incision and polymerase activities in determining the distribution of damaged sites that are mended in xeroderma pigmentosum group C cells. Cancer Res 47: 2393–2396

Cleaver JE, Thomas GH (1988) Rapid diagnosis of sensitivity to ultraviolet light in fibroblasts from dermatologic disorders, with particular reference to xeroderma pigmentosum. J Invest Derm 90: 467–471

Cleaver JE, Kaufmann WK, Kapp LN, Park SD (1983) Replicon size and excision repair as factors in the inhibition and recovery of DNA synthesis from ultraviolet damage. Biochim Biophys Acta 739: 207–215

Cleaver JE, Cortes F, Lutxe LH, Morgan WF, Player AN, Mitchell DL (1987) Unique DNA repair properties of a xeroderma pigmentosum revertant. Mol Cell Biol 7: 3353–3357

Cohn SM, Lieberman MW (1984) The use of antibodies to 5-bromo-2'-deoxyuridine for the isolation of sequences containing excision-repair sites. J Biol Chem 259: 12456–12462

Collins A, Downes CS, Johnson RT (1984) DNA repair and its inhibition. IRL Press, Oxford

Crawford BD, Enger MD, Griffith BB, Griffith JK, Hanners JL, Longmire JL, Munk AC, Stallings RL, Tesmer JG (1985) Coordinate amplification of metallothionein I and II genes in cadmium-resistant Chinese hamster cells. Mol Cell Biol 5: 320–329

Dijkwel PA, Hamlin JL (1988) Matrix attachment regions are positioned near replication initiation sites, genes, and an interamplicon junction in the amplified dihydrofolate reductase domain of CHO cells. Mol Cell Biol 8: 5398–5409

Dipple A, Rogers JJ (1977) Excision of 7-bromomethylbenz[a]anthracene-DNA adducts in replicating mammalian cells. Biochemistry 16: 1499–1503

Edwards S, Fielding S, Waters R (1987) The response to DNA damage induced by 4-nitroquinoline-1-oxide or its 3-methyl derivative in xeroderma pigmentosum fibroblasts belonging to different complementation groups: evidence for different epistasis groups. Carcinogenesis (Lond) 8: 1071–1075

Fornace AJJr, Schalch H, Alamo IJr (1988) Coordinate induction of metallothioneins I and II in rodent cells by UV-irradiation. Mol Cell Biol 8: 4716–4720

Friedberg EC (1984) DNA repair. Freeman, NY

Grossman L, Caron PR, Mazur SJ, Oh EY (1988) Repair of DNA containing pyrimidine dimers. FASEB J 2: 2696–2701

Hamer DH (1986) Metallothioneins. Annu Rev Biochem 55: 913–951

Hanawalt PC (1986) Intragenomic heterogeneity in DNA damage processing: potential implications for risk assessment. Basic Life Sci 38: 489–498

Hanawalt PC (1987) Preferential DNA repair in expressed genes. Environ Health Perspect 76: 9–14

Hanawalt PC, Mellon I, Scicchitano D, Spivak G (1989) Relationships between DNA repair and transcription in defined DNA sequences in mammalian cells. In: Lambert MW (ed) DNA repair mechanisms and their biological implications. Plenum, NY, pp 325–337

Haqq CM, Smith CA (1987) DNA repair in tissue specific genes in cultured mouse cells. In: Fielden EM, Fowler JF, Hendry JH, Scott D (eds) Radiation research, vol. 2. Taylor & Francis, Lond, pp 418–423

Harless J, Hewitt RR (1987) Intranuclear localization of UV-induced DNA repair in human VA-13 cells. Mutat Res 183: 177–184

Ho L, Bohr VA, Hanawalt PC (1989) Demethylation enhances the removal of pyrimidine dimers from the overall genome and from specific DNA sequences in Chinese hamster ovary cells. Mol Cell Biol 9: 1594–1603

Ikenaga M, Kakunaga T (1977) Excision of 4-nitroquinoline 1-oxide damage and transformation in mouse cells. Cancer Res 37: 3672–3678

Irwin RT, Wogan GN (1984) Quantitation of aflatoxin B1 adduction within the ribosomal RNA gene sequences of rat liver. Proc Natl Acad Sci USA 81: 664–668

Kaneko M, Cerutti PA (1980) Excision of *N*-acetoxy-2-acetylaminofluorene-induced DNA adducts from chromatin fractions of human fibroblasts. Cancer Res 40: 4313–4319

Kantor GK, Barsalou LS (1988) A beta-actin intron probe hybridizes preferentially to the repaired DNA in XP-C cells. J Cell Biochem (Suppl) 12A: 291

Kantor GJ, Elking CF (1988) Biological significance of domain-oriented DNA repair in xeroderma pigmentosum cells. Cancer Res 48: 844–849

Kantor GJ, Hull DR (1984) The rate of removal of pyrimidine dimers in quiescent cultures of normal human and xeroderma pigmentosum cells. Mutat Res 132: 21–31

Kantor GJ, Player AN (1986) A further definition of characteristics of DNA-excision repair in xeroderma pigmentosum complementation group A strains. Mutat Res 166: 79–88

Kantor GJ, Warner C, Hull DR (1977) The effect of ultraviolet light on arrested human diploid cell populations. Photochem Photobiol 25: 483–489

Karentz D, Cleaver JE (1986) Excision repair in xeroderma pigmentosum group C but not group D is clustered in a small fraction of the total genome. Mutat Res 165: 165–174

Kas E, Chasin LA (1987) Anchorage of the Chinese hamster dihydrofolate reductase gene to the nuclear scaffold occurs in an intragenic region. J Mol Biol 198: 677–692

Kessler O, Ben-Ishai R (1988) Lack of preferential DNA repair of a muscle specific gene during myogenesis. In: Friedberg EC, Hanawalt PC (eds) Mechanisms and consequences of DNA damage processing. Alan R Liss, Inc, NY, pp 267–272

Keyse SM, Tyrrell RM (1987) Rapidly occurring DNA excision repair events determine the biological expression of UV-induced damage in human cells. Carcinogenesis (Lond) 8: 1251–1256

Konze-Thomas B, Hazard RM, Maher VM, McCormick JJ (1982) Extent of excision repair before DNA synthesis determines the mutagenic but not the lethal effect of UV radiation. Mutat Res 94: 421–434

Leadon SA (1986) Differential repair of DNA damage in specific nucleotide sequences in monkey cells. Nucleic Acids Res 14: 8979–8995

Leadon SA (1988) Immunological probes for lesions and repair patches in DNA. In: Friedberg EC, Hanawalt PC (eds) DNA repair. A Laboratory Manual of Research Procedures. Marcel Dekker, Inc, NY, pp 311–326

Leadon SA, Snowden MM (1988) Differential repair of DNA damage in the human metallothionein gene family. Mol Cell Biol 8: 5331–5338

Leadon SA, Zolan ME, Hanawalt PC (1983) Restricted repair of aflatoxin B1 induced damage in alpha DNA of monkey cells. Nucleic Acids Res 11: 5675–5689

Linskens M, Huberman JA (1988) Organization of replication in the rDNA of *Saccharomyces cerevisiae*. Mol Cell Biol 8:4927–4935

Lippke JA, Gordon LK, Brash DE, Haseltine WA (1981) Distribution of light-induced damage in a defined sequence of human DNA: detection of alkali-sensitive lesions at pyrimidine nucleoside-cytosine sequences. Proc Natl Acad Sci, USA 78: 3388–3392

Looney JE, Ma C, Leu TZ, Flintoff WF, Troutman WB, Hamlin JL (1988) The dihydrofolate reductase amplicons in different methotrexate-resistant hamster cell lines share at least a 273-kb core sequence, but the amplicons in some cell lines are much larger and are remarkably uniform in structure. Mol Cell Biol 8: 5268–5279

Lusky KL, Faust JR, Chin DJ, Brown MS, Goldstein JL (1983) Amplification of the gene for 3-hydroxy-3-methylglutaryl coenzyme A reductase, but not for the 53-kDa protein, in UT-1 cells. J Biol Chem 258: 8462–8469

Madhani HD, Bohr VA, Hanawalt PC (1986) Differential DNA repair in transcriptionally active and inactive proto-oncogenes: c-*abl* and c-*mos*. Cell 45: 417–423

Maher VM, Dorney DJ, Mendrala AL, Konze-Thomas B, McCormick JJ (1979) DNA excision-repair processes in human cells can eliminate the cytotoxic and mutagenic consequences of ultraviolet light. Mutat Res 62: 311–323

Mansbridge JN, Hanawalt PC (1983) Domain-limited repair of DNA in ultraviolet irradiated fibroblasts from xeroderma pigmentosum complementation group C. In: Friedberg EC, Bridges BR (eds) Cellular responses to DNA damage. Alan R Liss, Inc, NY, pp 195–208

Matsumoto A, Vos JMH, Hanawalt PC (1989) Repair analysis of mitomycin C induced cross-linking in ribosomal RNA genes in lymphoblastoid cells from Fanconi's anemia patients. Mutat Res 217: 185–192

Mayne LV, Lehmann AR (1982) Failure of RNA synthesis to recover after UV irradiation: an early defect in cells from individuals with Cockayne's syndrome and xeroderma pigmentosum. Cancer Res 42: 1473–1478

Mayne LV, Mullenders LHF, Van Zeeland AA (1988) Cockayne's syndrome: An UV sensitive disorder with a defect in the repair of transcribing DNA but normal overall excision repair. In: Friedberg EC, Hanawalt PC (eds) Mechanisms and consequences of DNA damage processing. Alan R Liss, Inc, NY, pp 349–353

Mellon I, Bohr VA, Smith CA, Hanawalt PC (1986) Preferential DNA repair of an active gene in human cells. Proc Natl. Acad Sci USA 83: 8878–8882

Mellon I, Hanawalt PC (1989) Induction of the *E. coli* lactose operon selectively increases repair of its transcribed strand. Nature (in press)

Mellon I, Spivak G, Hanawalt PC (1987) Selective removal of transcription-blocking DNA damage from transcribed strand of the mammalian DHFR gene. Cell 51: 241–249

Mellon I, Spivak G, Hanawalt PC (1988) Strand specificity of DNA repair in CHO cells expressing the human *ERCC-1* gene. In: Friedberg EC, Hanawalt PC (eds) Mechanisms and consequences of DNA damage processing. Alan R Liss, Inc, NY, pp 263–266

Mitchell DL, Heipek CA, Clarkson JM (1985) (6-4)Photoproducts are removed from the DNA of UV-irradiated mammalian cells more efficiently than cyclobutane pyrimidine dimers. Mutat Res Lett 143: 109–112

Mitchell DL, Zdzienicka MZ, van Zeeland AA, Nairn R (1989) Intermediate (6-4) photoproduct repair in Chinese hamster VJ9 mutant V-H1 correlates with intermediate levels of DNA incision and repair replication Mutat Res 226: 43–47

Mitchell PJ, Carothers AM, Han JH, Harding JD, Kas E, Venolia L, Chasin LA (1986) Multiple transcription start sites, DNase I hypersensitive sites, and an opposite-strand exon in the 5' region of the CHO DHFR gene. Mol Cell Biol 6: 425–440

Montoya-Zavala M, Hamlin JL (1985) Similar 150 kilobase DNA sequences are amplified in independently derived methotrexate-resistant Chinese hamster cells. Mol Cell Biol 5: 619–627

Moustacchi E, Ehmann UK, Friedberg EC (1979) Defective recovery of semiconservative DNA synthesis in xeroderma pigmentosum cells following split-dose ultraviolet irradiation. Mutat Res 62: 159–171

Mullenders LH, Van Zeeland AA, Natarajan AT (1983) Analysis of the distribution of DNA repair patches in the DNA-nuclear matrix complex from human cells. Biochim Biophys Acta 740: 428–435

Mullenders LH, Van Kesteren AC, Bussmann CJ, Van Zeeland AA, Natarajan AT (1984) Preferential repair of nuclear matrix associated DNA in xeroderma pigmentosum complementation group C. Mutat Res 141: 75–82

Mullenders LH, Van Kesteren AC, Bussmann CJ, Van Zeeland AA, Natarajan AT (1986) Distribution of UV-induced repair events in higher-order chromatin loops in human and hamster fibroblasts. Carcinogenesis (Lond) 7: 995–1002

Mullenders LHF, Van Kesteren-Van Leeuwen AC, Van Zeeland AA, Natarajan AT (1988) Nuclear matrix associated DNA is preferentially repaired in normal human fibroblasts, exposed to a low dose of UV light, but not in Cockayne's syndrome fibroblasts. Nucleic Acids Res 16: 10607–10622

Nakanishi M, Goldstein JL, Brown MS (1988) Multivalent control of 3-hydroxy-3-methylglutaryl coenzyme A reductase. Mevalonate-derived product inhibits translation of mRNA and accelerates degradation of the enzyme. J Biol Chem 263: 8929–8937

Nose K, Nikaido O (1984) Transcriptionally active and inactive genes are similarly modified by chemical carcinogens and X-ray in normal human fibroblasts. Biochim Biophys Acta 781: 273–278

Okumoto DS, Bohr VA (1987) DNA repair in the metallothionein gene increases with transcriptional activation. Nucleic Acids Res 15: 10021–10030

Patterson MC, Middlestadt MV, MacFarlane SJ, Gentner NE, Weinfeld M, Eker APM (1987) Molecular evidence for cleavage of intradimer phosphodiester linkage as a novel step in excision repair of cyclobutyl pyrimidine photodimers in cultured human cells. J Cell Sci (Suppl) 6: 161–176

Peleg L, Raz E, Ben-Ishai R (1977) Changing capacity for DNA excision repair in mouse embryonic cells in vitro. Exp Cell Res 104: 301–307

Pirsel M, DiPaolo JA, Doniger J (1989) Transient appearance of photolyase-induced break sensitive sites in the DNA of ultraviolet irradiated Syrian hamster fetal cells. Mutat Res 217: 39–45

Player AN, Kantor GJ (1987) The endogenous nuclease sensitivity of repaired DNA in human fibroblasts. Mutat Res 184: 169–178

Poirier MC, Dubin MA, Yuspa SH (1979) Formation and removal of specific acetylaminofluorene-DNA adducts in mouse and human cells measured by radioimmunoassay. Cancer Res 39: 1377–1381

Rajagopalan S, Cohn SM, Lieberman ML (1984) UV radiation induced damage and repair in human ribosomal DNA sequences. Fed Proc 43: 1641

Ross PM, Yu HS (1988) Interstrand crosslinks due to 4,5',8-trimethylpsoralen and near ultraviolet light in specific sequences of animal DNA. Effect of constitutive chromatin structure and induced transcription. J Mol Biol 201: 339–351

Ryan AJ, Billett MA, O'Connor PJ (1986) Selective repair of methylated purines in regions of chromatin DNA. Carcinogenesis (Lond) 7: 1497–1503

Sancar A, Sancar GW (1988) DNA repair enzymes, Annu Rev Biochem 57: 29–67

Scicchitano DS, Hanawalt PC (1989) Repair of N-methylpurines in specific DNA sequences in Chinese hamster ovary cells: absence of strand specificity in the dihydrofolate reductase gene. Proc Natl Acad Sci USA 86: 3050–3054

Shi YB, Gamper H, Hearst JE (1987) The effects of covalent additions of a psoralen on transcription by *E. coli* RNA polymerase. Nucleic Acids Res 15: 6843–6854

Shi YB, Gamper H, Hearst JE (1988) Interaction of T7 RNA polymerase with DNA in an elongation complex arrested at a specific psoralen site. J Biol Chem 263: 527–534

Smith CA (1987) DNA repair in specific sequences in mammalian cells. J Cell Sci (Suppl) 6: 225–241

Smith CA (1988a) Discussion summary, fine structure of DNA repair. In: Friedberg EC, Hanawalt PC (eds) Mechanisms and consequences of DNA damage processing. Alan R Liss, Inc, NY, pp 397–404

Smith CA (1988b) Repair of DNA containing furocoumarin adducts. In: Gasparro FP (ed) Psoralen DNA photobiology, vol II. CRC Press, Boca Raton, pp 87–116

Szafarz D, Zajdela F, Bornecque C, Barat N (1983) Evaluation of DNA crosslinks and monoadducts in mouse embryo fibroblasts after treatment with mono- and bifunctional furocoumarins and 365 mn (UVA) radiation. Possible relationship to carcinogenicity. Photochem Photobiol 38: 557–562

Tang M, Bohr VA, Zhang X, Pierce J, Hanawalt PC (1989) Quantification of aminofmorene adduct formation and repair in defined DNA sequences in mammalian cells using the UVR-ABC nuclease. J Biol Chem 264: 14455–14462

Thomas DC, Morton AG, Bohr VA, Sancar A (1988) General method for quantifying base adducts in specific mammalian genes. Proc Natl Acad Sci USA 85: 3723–3727

Thompson LH, Brookman KW, Mooney CL (1984) Repair of DNA adducts in asynchronous CHO cells and the role of repair in cell killing and mutation induction in synchronous cells treated with 7-bromomethylbenz[a]anthracene. Somatic Cell Mol Genet 10: 183–194

Thompson LH, Weber CA, Jones NJ (1989) Human DNA repair and recombination genes. In: Lambert MW (ed) DNA repair mechanisms and their biological implications in mammalian cells. Plenum, NY, pp 547–561

Tyrell RM, Amaudruz F (1987) Evidence for two independent pathways of biologically effective excision repair from its rate and extent in cells cultured from sun-sensitive humans. Cancer Res 47: 3725–3728

Van Duin M, Janssen JH, De Wit J, Hoeijmakers JHJ, Thompson LH, Bootsma D, Westerveld A (1988) Transfection of the cloned human excision repair gene *ERCC-1* to UV-sensitive CHO mutants only corrects the repair defect in complementation group-2 mutants. Mutat Res 193: 123–130

Van Zeeland AA, Smith CA, Hanawalt PC (1981) Sensitive determination of pyrimidine dimers in DNA of UV-irradiated mammalian cells. Introduction of T4 endonuclease V into frozen and thawed cells. Mutat Res 82: 173–189

Vos J-M (1988) Analysis of psoralen monoadducts and interstrand crosslinks in defined genomic sequences. In: Friedberg EC, Hanawalt PC (eds) DNA repair. A Laboratory Manual of Research Procedures. Marcel Dekker, Inc, NY, pp 367–398

Vos J-M, Hanawalt PC (1987) Processing of psoralen adducts in an active human gene: repair and replication of DNA containing monoadducts and interstrand cross-links. Cell 50: 789–799

Vrieling H, Simons JWIM, Van Zeeland AA (1988) Nucleotide sequence determination of point mutations at the mouse HPRT locus using in vitro amplification of HPRT mRNA sequences. Mutat Res 198: 107–114

Vrieling H, Van Rooyen ML, Groen NA, Zdzienicka, Simons JWIM, Lohman PHM, Van Zeeland AA (1989) DNA strand specificity for UV-induced mutations in mammalian cells. Mol Cell Biol 9: 1277–1283

Vuksanovic L, Cleaver JE (1987) Unique cross-link and monoadduct repair characteristics of a xeroderma pigmentosum revertant cell line. Mutat Res 184: 255– 263

Zdzienicka MZ, Van der Schans GP, Westerveld A, Van Zeeland AA, Simons JWIM (1988) Phenotypic heterogeneity within the first complementation group of UV-sensitive mutants of Chinese hamster cell lines. Mutat Res 193: 31–41

The Use of Short-Term Genotoxicity Tests in Risk Assessment

F.K. ENNEVER[1]

Contents

1 Introduction

The discipline of risk assessment is currently under pressure from two directions. On the one hand, more and more decisions must be made concerning how resources are to be allocated to reduce human exposure to harmful chemicals, and the assessment of risk from those exposures is often essential in making such decisions. On the other hand, many have emphasized the gaps in scientific knowledge and the reliance on assumptions in risk assessments, and the resulting regulations have been criticized as too lenient by some (Epstein and Swartz 1988; Karstadt 1988; Perera 1988; Bailar et al. 1988) and as too stringent by others (Wilson 1988; Freedman and Zeisel 1988). Risk assessments will always have some component of uncertainty, because we often care about risk at levels well below what is detectable [for example, many environmental regulations strive for a lifetime increased cancer risk of no more than 10^{-5} or 10^{-6} (Anderson 1988), but the lowest detectable risk in humans is about 10^{-3}, the risk of diethylstilbestrol exposure, which causes a very rare form of tumor (IARC 1987)], and so some extrapolation will always be necessary. Also, because generating information requires the expenditure of money, the ideal risk assessment with perfect information can be approached only for a very limited number of chemicals. Thus, progress in risk assessment will come not only from identifying areas of uncertainty where more research is needed, but also from developing rational procedures for making decisions in the presence of uncertainty.

Traditionally, the assessment of the risk of chemically-caused cancers has been concerned with procedures for extrapolating effects observed in experimental

[1] Department of Environmental Health Sciences, School of Medicine, Case Western Reserve University, Cleveland, OH 44106, USA

carcinogenesis bioassays (lifetime exposure of rodents to a chemical) to effects expected in humans at much lower doses (Anderson 1988). However, the rodent cancer bioassay is so expensive (more than $ 1 million per chemical) that the number of substances tested will always remain quite limited. Approximately 1000 chemicals have been tested in some form of the rodent cancer bioassay, and about 600 are positive, 100 are negative, and the remaining 300 are inconclusive (no effect was produced, but the protocol was not adequate to conclude noncarcinogenicity) (Gold et al. 1984, 1986, 1987). However, there are a total of about 60 000 existing chemicals in commerce and industry, and about 1000 new chemicals are introduced each year (NAS 1984); and these estimates do not include most naturally-occurring products that humans ingest in food and beverages (Ames et al. 1987). Thus, unless the selection of the 1000 chemicals tested in rodents has been extraordinarily lucky, the carcinogens that we know about may be far outnumbered by the carcinogens we do not know about.

Short-term tests are substantially less expensive and time-consuming to perform than the rodent cancer bioassay; however, they do not measure cancer induction but rather a variety of biological activities related to cancer causation: DNA damage, mutation, chromosomal damage, interference with cell replication, and neoplastic transformation. (For convenience, the term „genotoxicity" will be used here to indicate the variety of end points measured by short-term tests.) In traditional risk assessments the role of short-term tests has generally been limited to suggesting that nongenotoxic carcinogens (defined in a variety of ways) may have a mechanism of action which exhibits a threshold (Flamm and Scheuplein 1988). This chapter will outline several areas in which short-term tests may have a key role in minimizing chemically caused cancer in humans, the ultimate goal of risk assessment.

2 Applications of Short-Term Tests

2.1 The Prevalence of Carcinogens in the Chemical Universe

All strategies for classifying chemicals as human carcinogens or noncarcinogens are prone to both false negatives (a chemical actually carcinogenic to humans but treated as noncarcinogenic) and false positives (a chemical actually not carcinogenic to humans but treated as carcinogenic). The concepts of false positives and false negatives are related to sensitivity (the probability that a strategy will give a positive result for a carcinogen) and specificity (the probability that a strategy will give a negative result of a noncarcinogen). One extreme method of classification is to treat all chemicals as noncarcinogens, as existing chemicals generally are treated, which has perfect specificity (there will be no false positives), but zero sensitivity (the number of false negatives will be equal to the proportion of carcinogens actually in the population). Another extreme method is to treat all chemicals as carcinogens, which has perfect sensitivity (there will be no false negatives), but zero specificity (the number of false positives will be equal to the proportion of noncarcinogens actually in the population). An intermediate strategy is to classify chemicals using one or more tests (structure-activity relationships, short-term genotoxicity tests, and/or the rodent cancer bioassay), calling chemicals carcinogens if they are positive in the test and noncarcinogens if they are negative in the test. The number of false positives and negatives resulting from a testing strategy depends on the inherent accuracy of the test or battery (test sensitivity and specificity), but it also depends on the underlying proportion of carcinogens in the population, because the more carcinogens in the population, the more false negatives there will be for a given sensitivity, and the more noncarcinogens in the population, the more false positives there will be for a given specificity (Lave and Omenn 1986; Ennever and Rosenkranz 1988a; Lave et al. 1988).

In many cases, trade-offs can be made between the sensitivity and specificity of a testing strategy; for example, a single test can be made more sensitive by relaxing the criteria for a positive result (e.g., requiring a doubling rather than a tripling of spontaneous mutation frequency), or a battery of tests could be made more specific requiring a majority of results rather than a single result to be positive in order to classify the overall response as positive. Deciding upon an optimal strategy involves balancing false positives and false negatives, and therefore requires knowledge of the underlying proportion of carcinogens in the chemical population (Lave and Omenn 1986; Ennever and Rosenkranz 1988a; Lave et al. 1988). For example, for representative batteries of three short-term tests, the optimal strategy changes from requiring three negatives to predict noncarcinogenicity to requiring two negatives to requiring only one negative as the underlying proportion of carcinogens goes from 50 to 20 to 5% (Ennever and Rosenkranz 1988a).

The prevalence of carcinogens in the chemical universe is not easy to estimate directly, because of the high cost of the rodent cancer bioassay. However, the proportion of mutagens may be calculated, and provides an indirect estimate of the proportion of carcinogens (Ennever and Rosenkranz 1988b). For a given assay system i, the proportion of chemicals in a defined population that are active in the assay is symbolized A_i^+. (Known carcinogens and noncarcinogens are excluded from the population, because the analysis requires considering the population to be composed entirely of carcinogens or entirely of noncarcinogens, see below). The sensitivity of the assay, the probability that the assay i will give a positive result for a carcinogen, is symbolized a_i^+ , and the specificity of the assay, the probability that the assay will give a negative result for a noncarcinogen, is symbolized a_i^-. When the assay is applied to a set of chemical with an underlying proportion of carcinogens C^+, the number of positive results A^+ is given by:

$$A^+ = (a^+)(C^+) + (1 - a^-)(1 - C^+) \quad (1)$$

and the number of negative results $1 - A^+$ is:

$$1 - A^+ = (1 - a^+)(C^+) + (a^-)(1 - C^+). \quad (2)$$

A^+ is known, but C^+, a^+, and a^- are not, and so Eqs. (1) and (2) do not have a unique solution. However, it is possible to place limits on C^+ and a^+ if A^+ is large, or on C^+ and a^- if A^+ is small (Ennever and Rosenkranz 1988b).

The most widely used short-term genotoxicity test is the *Salmonella typhimurium* reverse mutation assay (abbreviated here as Sty). Two distinct data bases are available for this assay; from the Gene-Tox program of the US Environmental Protection Agency, a peer-reviewed compilation of results from the published literature (Kier et al. 1986), and from the National Toxicology Program (NTP), original testing done under code using a standard protocol (Haworth et al. 1983; Dunkel et al. 1984, 1985; Mortelmans et al. 1986; Zeiger et al. 1987, 1988). The Gene-Tox data base will be analyzed first. Excluding known carcinogens and noncarcinogens, results are reported for 872 chemicals: 109 were nonmutagenic, 297 were questionable, and 466 were mutagenic. We will treat questionable results as "no test" and remove them from the data base before analysis. Thus, A_G^+, the proportion of positive results in the Sty Gene-Tox data base, is 109/(109 + 466) = 0.810. The minimum possible value for C^+ is 0.810 (in which case a^+ and a^- must both be 1.0), and the minimum possible value for a^+ is also 0.810 (in which case C^+ must be 1.0). Only two assumptions are required to make these inferences for Sty or any other test:

1. There is *some* relationship between carcinogenicity and the genotoxic end point measured by the test. If there is no relationship, then the test gives the same proportion of positive results for both carcinogens and noncarcinogens (in mathematical terms, $a^+ = 1 - a^-$), or gives fewer positive results for carcinogens than noncarcinogens ($a^+ < 1 - a^-$). If the test gives an even slightly greater proportion of positive results for carcinogens than noncarcinogens ($a^+ > 1 - a^-$), then the analysis holds; the test might not be very predictive if the specificity is only slightly greater than one minus the sensitivity, but the bound on C^+ is still valid.

2. The population of chemicals was not selected on the basis of the test results [e.g., if every chemical giving a positive response in the assay were eliminated from the data set ($A^+ = 0.0$), the conclusion that $C^+ = 0.0$ and $a^- = 1.0$ would not be valid because a^- would not be a true estimate of probability for the given population of chemicals].

The first assumption seems reasonable for the Sty test and most other genotoxicity assays. The second assumption may be somewhat more tenuous for the Gene-Tox data base, because although the selection for inclusion in the data base did not depend upon whether the result was positive or negative, it did depend on the result being published, and there often has been a bias against publishing negative results. Quantifying the effect of such an indirect selection pressure is difficult; since the bias against negative results may be lessening, some insight may be gained by comparing the planned update of the Gene-Tox Sty data base which will include data published more recently.

The other source of Sty data, from the NTP, does fulfill the second assumption, because all chemicals for a given publication were selected prior to the start of testing, and all results were reported. Excluding known carcinogens and noncarcinogens, the NTP data base contains results for 883 chemicals: 620 were nonmutagenic, 18 were questionable, and 245 were mutagenic. Again removing questionable results, A_N^+, the proportion of positive results in the Sty NTP data base, is 245/(620 + 245) = 0.283. The maximum possible value for C^+ is 0.283 (in which case a^+ and a^- must both be 1.0), and the minimum possible value for a^- is 1 – 0.283 = 0.717 (in which case C^+ must be 0.0).

A further analysis can be made of the two Sty data bases by removing from consideration chemicals present in both Gene-Tox and NTP data bases, because such chemicals cannot be both all carcinogens and all noncarcinogens. A total of 52 chemicals of unknown carcinogenicity had unequivocal results in both data bases: 21 were negative in both, 23 were positive in both, 4 were positive in Gene-Tox and negative in NTP, and 4 were negative in Gene-Tox and positive in NTP. Revised limits for A_G^+ and A_N^+ are 0.839 and 0.268, respectively, giving revised limits on a^+ of 0.839 and on a^- of 0.732.

Sensitivity and specificity have been defined here as probabilities, sensitivity as the probability that Sty will give a positive result for a carcinogen, and specificity as the probability that Sty will give a negative result for a noncarcinogen. The usual way that sensitivity and specificity are estimated is by calculating the proportion of known carcinogens giving positive results and the proportion of known noncarcinogens giving negative results (Chankong et al. 1985; Tennant et al. 1987), but such a procedure depends on the assumption that the chemicals known to be carcinogenic and noncarcinogenic are a representative random sample of all carcinogens and noncarcinogens. In the Gene-Tox data base, Sty was positive for 128 of 157 carcinogens tested, a nominal sensitivity of 0.815, and was negative for 10 of 16 noncarcinogens tested, a nominal specificity of 0.625. In the NTP data base, Sty was positive for 91 of 142 carcinogens tested, a nominal sensitivity of 0.642, and was negative for 43 of 67 noncarcinogens tested, a nominal specificity of 0.641. Combining the data bases yields a nominal sensitivity of 0.740 and a nominal specificity of 0.623. The limits developed by the indirect method described above were based on test results for many more chemicals than the direct calculation, and so may be more representative of the performance of the Sty assay on large groups of chemicals. The difference in the values may also indicate that the chemicals which have been tested in the rodent bioassay are not entirely representative of carcinogens and noncarcinogens in the chemical universe.

The method of analysis described here gives limits not only on sensitivity and specificity, but also on the proportion of carcinogens in the population of chemicals tested. At least 84% of the 523 chemicals in the Gene-Tox data base must be carcinogenic, and 100% would be if the sensitivity of Sty were as low as 0.84. On the other hand, in the NTP data base, no more than 27% of the 813 chemicals can be carcinogenic, and none are carcinogenic if the specificity of Sty is as low as 0.73. Neither data base was selected to be representative of chemicals as a whole, and so neither 84% or 27% should be used as an estimate of the prevalence of carcinogens in the chemical universe. Preliminary reports of results of systematic testing of chemicals in Japan indicate that no more than 12% of chemicals are positive in sty (Matsushima, 1987), which would imply that no more than 12 % of chemicals are carcinogenic, and fewer than 12% unless the specificity of Sty were perfect. Further analysis of the Japanese experience and perhaps of testing results of industrial concerns is planned so as to continue analyzing short-term test results to estimate the prevalence of carcinogens in the chemical universe.

2.2 The Role of Metabolism in Determining Biological Activity

One issue in risk assessment which has received increasing attention is the relatively low concordance between carcinogenicity in mice and in rats (Tennant et al. 1987; Ashby and Tennant 1988; Brockman and DeMarini 1988). The significance for human risk of a chemical which is carcinogenic to rats but not to mice, or vice versa, is difficult to evaluate without knowing the reasons for the differing results. One possibility is that differences in metabolism (transformation of the parent chemical into both active and inactive metbolites) between rats and mice cause the differences in carcinogenicity, but directly evaluating this hypothesis requires a research effort to identify, for individual chemicals, the active metabolite and the activation, deactivation, and excretion pathways in each species.

Again, short-term tests can be used as an indirect measure of the importance of metabolism in species-specific carcinogenicity. This analysis used a data base of 73 chemicals which had been tested by the National Toxicology Program (Tennant et al. 1987), using standard protocols for carcinogenicity and for genotoxicity in four systems: *Salmonella* mutagenicity (Sty), chromosomal aberrations in Chinese hamster ovary cells in vitro (Cvt), sister chromatid exchanges in Chinese hamster ovary cells (SCE), and mutations in mouse lymphoma cell line L5178Y (Mly). Carcinogenicity results for the 73 chemicals were classified into six categories following the scheme of Ashby and Tennant (1988): A = both rats and mice affected; B = a single species with two or more tissues affected; C = a single species with a single tissue affected; D = a single sex of a single species with a single tissue affected; E = an adequate study providing equivocal evidence of carcinogenicity; and F = noncarcinogenic.

Genotoxicity tests were routinely performed in both the presence and absence of exogenous activation (S9 microsomal fractions prepared by a standard protocol). Four categories of results were reported (Tennant et al. 1987): N,A = active both in the prescence and absence of exogenous activation; N = active only in the absence of exogenous activation; A = active only in the presence of exogenous activation; and – = inactive. Table 1 tabulates the categories of genotoxicity responses for Sty, Cvt, and SCE for the six carcinogenicity categories. (The Mly assay was often not performed with exogenous activation if the chemical was active without it, and so a similar analysis could not be done.) Of the ten chemicals mutagenic in Sty both with and without activation, seven were group A, that is, trans-species carcinogens. This association is marginally statistically significant if the proportion of N,A results in all positive results is compared for group A versus groups B through F [7/11 vs 3/13, $p < 0.11$ by the ratio test, Fleiss (1981)], and highly significant if the proportion of N,A results in all results is compared (7/20 vs 3/53, $p < 0.004$).

The test Cvt also shows an association between group A carcinogens and group N,A clastogens (Table 1): of the 17 chemicals clastogenic both with and without activation, 9 were group A, 2 each were groups B, C, and D, and 1 each were groups E and F. Again, the association is marginally significant for N,A responses among positive responses, and significant for N,A responses among all responses ($p < 0.1$ and $p < 0.02$, respectively). However, SCE shows no clear

Table 1. Comparison of activity in genotoxicity assays and carcinogenicity for 73 chemicals from Tennant et al. (1987)[a]

Carcinogenicity	Sty				Cvt				SCE				
	N,A	N	A	–	N,A	N	A	–	N,A	N	A	–	total
A	7	0	4	9	9	1	2	8	9	5	2	4	20
B	1	1	3	4	2	1	2	4	5	2	0	2	9
C	0	1	1	5	2	2	0	3	3	3	0	1	7
D	1	0	1	6	2	0	1	5	0	1	2	5	8
E	0	0	0	9	1	1	2	5	4	1	1	3	9
F	1	0	3	16	1	1	3	15	8	1	1	10	20
	10	2	12	49	17	6	10	40	29	13	6	25	73

[a] Test abbreviations: Sty = *Salmonella* mutagenicity; Cvt = chromosomal aberrations in Chinese hamster ovary cells in vitro; SCE = sister chromatid exchanges in Chinese hamster ovary cells. Carcinogenicity categories: A = both rats and mice affected; B = a single species with two or more tissues affected; C = a single species with a single tissue affected; D = a single sex of a single species with a single tissue affected; E = an adequate study providing equivocal evidence of carcinogenicity; and F = noncarcinogenic (Ashby and Tennant 1988). Genotoxicity categories: N,A = active both in the presence and in the absence of exogenous activation; N = active only in the absence of exogenous activation; A = active only in the presence of exogenous activation; and – = inactive (Tennant et al. 1987).

association (Table 1): of the 29 chemicals clastogenic with and without activation, 9 were group A, 5 were group B, 3 were group C, 4 were group E, and 8 were group F. As mentioned above, in the Mly assay, a chemical was often never tested in the presence of exogenous activation if positive without it. As a consequence, only three chemicals were shown to be positive both with and without exogenous activation in Mly. Two were group A, but because of the small number of chemicals and the possibility that chemicals positive without activation could also have been positive if activation had been added, no firm conclusions can be drawn from the data.

A chemical can be positive in a short-term test system without exogenous metabolic activation either because it requires no metabolic transformation to be active or because it is activated by the inherent metabolism of the test organism. If the chemical is also positive in the presence of exogenous activation, this indicates that the genotoxicity is stable enough not to be abolished when proteins are added with which the chemical could react (e.g., with sulfhydryl groups) or when enzymes are added which could metabolize the chemical to an inactive form. Of course, an alternative possibility is that the chemical is transformed into a genotoxic form both by the inherent metabolism of the test organism and also by exogenous activation. Although the exogenous activation used in genotoxicity testing may not mimic in vivo metabolism (Douglas et al. 1988), genotoxicity which is manifest both in the absence and presence of exogenous activation is clearly a more robust activity than genotoxicity only in the absence or only in the presence of activation. Table 1 shows that for Sty and Cvt, chemicals with this robust genotoxicity tend to be trans-species carcinogens, a category of rodent carcinogens suggested to be more likely to be relevant to human risk (Ashby and Tennant 1988).

Salmonella cells are known to be capable of a variety of metabolic transformations, such as nitroreduction and anaerobic metabolism (Howard et al. 1987; McCoy et al. 1983). The Chinese hamster ovary (CHO) cells used for Cvt are generally thought to be deficient in microsome-mediated metabolic capability, but only two chemicals were positive only without activation in Sty, while seven were positive only without activation in Cvt (see Table 1). A third test system, sister chromatid exchanges (SCE), also using CHO cells, did not show the association between robust genotoxicity and trans-species carcinogenicity; that is, the chemicals causing sister chromatid exchanges both with and without activation were as likely to be single-sex, single-species, single-site carcinogens or noncarcinogens as they were to be trans-species carcinogens. Since Cvt and SCE were performed in the same cell line, the difference is probably not due to differing inherent metabolism (including uptake, absorption, and distribution), but to different end points and the mechanism which they represent. Of course, it could also reflect details of the protocol, such as treatment times, addition of bromodeoxyuridine, or number of cells scored (Galloway et al. 1985).

The results presented in Table 1 indicate that for Sty and Cvt, chemicals active under two different sets of metabolic conditions were primarily carcinogenic to both rats and mice, and chemicals active only with or only without metabolic activation were primarily carcinogenic to a single species, sex, or tissue or were noncarcinogenic. Thus, chemicals which are carcinogenic only to a single species, sex, or tissue may be active only under very specific metabolic conditions, whereas trans-species carcinogens may be active under a variety of metabolic conditions. The existence of single-species carcinogens has raised questions about the extrapolation of rodent carcinogenicity data in the assessment of human risk (Brockman and DeMarini 1988; Heddle 1988; Lave et al. 1988). These preliminary results indicate that differing metabolism may be responsible for a large part of the lack of concordance between rats and mice. If so, then an assessment of comparative metabolism in rats, mice, and humans could contribute to a reduction in uncertainty in risk analyses of single-species carcinogens.

In addition, these results, particularly if confirmed by analyses of larger data bases, may enhance the role of short-term genotoxicity tests in hazard evalution. Such tests are now interpreted to indicate the presence of a potential hazard, but not its magnitude. If the association between activity in both the presence and absence of exogenous activation and trans-species carcinogenicity is validated, then chemicals most likely to pose a threat to human health can be identified as those positive both with and without activation in certain short-term tests. This would assist in prioritizing existing chemicals for further investigation and in evaluating the potential hazard of new chemicals using short-term tests as a cost-effective screen.

2.3 Defining Genotoxicity and Nongenotoxicity

As mentioned above, the use of short-term tests in risk assessment has generally been confined to categorizing a known carcinogen as genotoxic or nongenotoxic. The relevant question is whether the mechanism of action of a carcinogen involves genotoxicity, both in the species showing a positive response, and ultimately also in the target tissue(s) of the human body at low doses. However, since answering either of those questions is a research project, and since there are a variety of short-term tests which measure genotoxicity as their end points, much of the discussion on defining genotoxicity and nongenotoxicity has concerned interpreting the results of short-term tests (Perera 1984; Ashby 1988).

The simplest way to define genotoxicity and nongenotoxicity is to use a single assay such as Sty (Ashby and Tennant 1988) or Mly (Clive 1988) or to use a structural scheme that identifies electrophilic substituents (Ashby and Tennant 1988). No decision rules are needed; a chemical is labeled genotoxic if it is positive in the assay and nongenotoxic if it is negative, or is labeled genotoxic if it has an electrophilic substituent, and nongenotoxic otherwise. More complex situations arise when more than one test is to be used. One type of decision rule counts positive and negative results, setting criteria for how many or what proportion of positive results are needed to call a chemical genotoxic (Perera 1984; Tennant et al. 1987; Green 1988). Other decision rules would weigh tests in a variety of ways: relevance to human genotoxicity (Brusick et al. 1986; Clayson 1987; ICPEMC 1988); known performance (sensitivity and specificity) on carcinogens and noncarcinogens (Rosenkranz and Ennever 1988) or electrophiles and nonelectrophiles (Ashby 1988); or predisposition to false positives for technical or mechanistic reasons (Ashby 1988; Green 1988).

From the perspective of risk assessment, classification of a chemical as genotoxic or nongenotoxic must include policy as well as scientific dimensions. When a classification of nongenotoxicity influences a risk assessment, the effect is almost always less stringent regulation, because the effects of nongenotoxic carcinogens may be assumed to have a threshold or be reversible, or for some other reason pose less of a hazard to humans than do genotoxic chemicals (Moore 1988; Wilson 1988; Perera 1988). Thus, any method of interpreting genotoxicity data implictly or explicitly contains trade-offs between classifying truly nongenotoxic chemicals as genotoxic and classifying truly genotoxic chemicals as nongenotoxic. The resolution of this trade-off should depend upon the consequences of false classifications. If many new chemicals are to be screened as candidates for a consumer product, a chemical could be called genotoxic on the strength of any positive short-term test result, no matter how many negative results exist. At the other extreme, there are many chemicals with extensive human exposure [e.g., 2,3,7,8–tetrachlorodibenzo-p-dioxin and di(2-ethylhexyl)phthalate] which should have their mechanisms of action investigated intensively (Clayson 1989), because of the potential consequences of a classification as nongenotoxic.

One promising direction for improving strategies for defining genotoxicity and nongenotoxicity is changes in the rodent cancer bioassay to include determination of whether genotoxicity (such as activation of an oncogene) is involved in the mechanism of carcinogenicity (Reynolds et al. 1987). Other approaches involve investigating the genotoxicity of human carcinogens and probable human noncarcinogens (Ennever et al. 1987; Shelby 1988). However, the scientific basis for genotoxicity classifications will never be perfect, and yet such classifications will continue to be needed. Thus, policy questions will always be present, and procedures should be developed for balancing potential errors resulting from uncertainty, taking into account the consequences of calling a chemical genotoxic or nongenotoxic.

3 Conclusions

Risk assessment and the use of short-term tests appeared simpler 15 years ago, when carcinogenicity appeared to be a rare property of chemicals, and the paradigm that "carcinogens are mutagens" (Ames et al. 1973) had few counterexamples. Greater understanding of the carcinogenesis process and the role of mutational activation of oncogenes (Stowers et al. 1987; Reynolds et al. 1987) has also been accompanied by much greater complexity in the measurement of biological activity. We now know of carcinogens that are nongenotoxic and noncarcinogens that are genotoxic, and of carcinogens with specific actions such as causing only benign tumors, causing tumors only in organs with high spontaneous rates, causing tumors only at doses which alter hormonal, metabolic, or other physiologic processes, or causing tumors only in a single tissue of a single sex of a single species (Tennant et al. 1987; Ashby and Tennant 1988). So far, no universal mechanism has appeared to explain any of these categories.

The ultimate question to be answered by risk assessment is "How should society allocate resources toward reducing chemically caused cancers?" Resources include money spent to clean up hazardous chemicals, to find substitutes or otherwise control exposure to carcinogens, and to monitor and enforce existing standards. However, resources also include generation of information needed to prioritize attention given to individual chemical substances. Short-term tests are rapid and relatively inexpensive, and can make a substantial contribution to risk assessment. Three areas described in this chapter, estimating the proportion of carcinogens in the chemical universe, identifying chemicals most likely to be trans-species carcinogens, and classifying carcinogens as genotoxic or nongenotoxic, are only a sample of the roles which short-term tests can play in helping to protect humans from chemical carcinogens.

Acknowledgements. I thank L.B. Lave, G.S. Omenn, and H.S. Rosenkranz for helpful discussions.

References

Ames BN, Durston WE, Yamasaki E, Lee FD (1973) Carcinogens are mutagens: a simple test system combining liver homogenates for activation and bacteria for detection. Proc Nat Acad Sci USA 70: 2281–2285

Ames BN, Magaw R, Gold LS (1987) Ranking possible carcinogenic hazards. Science 236: 271–280

Anderson EL (1988) The risk analysis process. In: Travis CC (ed) Carcinogen risk assessment. Plenum, New York, pp 3-17

Ashby J (1988) Computer assisted short-term test battery design: some questions. Environ Mol Mutagen 11: 443–448

Ashby J, Tennant RW (1988) Chemical structure, *Salmonella* mutagenicity and extent of carcinogenicity as indicators of genotoxic carcinogenesis among 222 chemicals tested in rodents by the U.S. NCI/NTP. Mutation Res 204: 17–115

Bailar JC III, Crouch EAC, Shaikh R, Spiegalman D (1988) One-hit models of carcinogenesis: conservative or not? Risk Anal 8: 485–497

Brockman HE, DeMarini DM (1988) Utility of shorts-term tests for genetic toxicity in the aftermath of the NTP's analysis of 73 chemicals. Environ Mol Mutagen 11: 421–435

Brusick D, Ashby J, de Serres F, Lohman P, Matsushima T, Matter B, Mendelsohn M, Waters M (1986) Weight-of-evidence scheme for evaluation and interpretation of short-term results. In: Ramel C, Lambert B, Magnusson J (eds) Genetic toxicology of environmental chemicals, Part B: Genetic effects and applied mutagenesis. Alan R Liss, Inc, New York, pp 121–129

Chankong V, Haimes YY, Rosenkranz HS, Pet-Edwards J (1985) The carcinogenicity prediction and battery selection (CPBS) method: a Bayesian approach. Mutation Res 153: 135–166

Clayson DB (1987) The need for biological risk assessment in reaching decisions about carcinogens. Mutation Res 185: 243–269

Clayson DB (1989) Can a mechanistic rationale be provided for non-genotoxic carcinogens identified in rodent bioassays? Mutation Res (in press)

Clive D (1988) Genetic toxicology: can we design predictive in vivo assays? Mutation Res 205: 313–330

Douglas GR, Blakey DH, Clayson DB (1988) Genotoxicity tests as predictors of carcinogens: an analysis (ICPEMC Working Paper No. 5). Mutation Res 196: 83–93

Dunkel V, Zeiger E, Brusick D, McCoy E, McGregor D, Mortelmans K, Rosenkranz HS, Simmon VF (1984) Reproducibility of microbial mutagenicity assays: I. Tests with *Salmonella typhimurium* und *Escherichia coli* using a standard protocol. Environ Mutagen 6 (Suppl 2): 1–251

Dunkel V, Zeiger E, Brusick D, McCoy E, McGregor D, Mortelmans K, Rosenkranz HS, Simmon VF (1985) Reproducibility of microbial mutagenicity assays: II. Testing of carcinogens and noncarcinogens in *Salmonella typhimurium* and *Escherichia coli*. Environ Mutagen 7 (Suppl 5): 1–248

Ennever FK, Rosenkranz HS (1988a) Influence of the proportion of carcinogens on the cost effectiveness of short-term tests. Mutation Res 197: 1–13

Ennever FK, Rosenkranz HS (1988b) Indirect estimates of the sensitivity and specificity of the *Salmonella* assay. Environ Mol Mutagen 11 (Suppl 11): 32 (Abstr)

Ennever FK, Noonan TJ, Rosenkranz HS (1987) The predictivity of animal bioassays and short-term genotoxicity tests for carcinogenicity and noncarcinogenicity to humans. Mutagenesis 2: 73–78

Epstein SS, Swartz JB (1988) Carcinogenic risk estimation. Science 240: 1043–1045

Flamm WG, Scheuplein RJ (1988) Use of short-term test data in risk analysis of chemical carcinogens. In: Travis CC (ed) Carcinogen risk assessment. Plenum, New York, pp 37–48

Fleiss JL (1981) Statistical methods for rates and proportions, J. Wiley & Sons, New York, 2nd Edn

Freedman DA, Zeisel H (1988) From mouse-to-man: the quantitative assessment of cancer risks. Statist Sci 3: 3–28, 45–56

Galloway SM, Bloom AD, Resnick M, Margolin BH, Nakamura F, Archer P, Zeiger E (1985) Development of a standard protocol for in vitro cytogenetic testing with Chinese hamster ovary cells: comparison of results for 22 compounds in two laboratories. Environ Mutagen 7: 1–51

Gold LS, Sawyer CB, Magaw R, Backman GM, de Veciana M, Levinson R, Hooper NK, Havender WR, Bernstein L, Peto R, Rike MC, Ames BN (1984) A carcinogenic potency database of the standardized results of animal bioassays. Environ Health Perspect 58: 9–319

Gold LS, de Veciana M, Backman GM, Magaw R, Lopipero R, Smith M, Blumenthal M, Levinson R, Bernstein L, Ames BN (1986) Chronological supplement to the carcinogenic potency database: standardized results of animal bioassays published through December 1982. Environ Health Perspect 67: 161–200

Gold LS, Slone TH, Backman GM, Magaw R, Da Costa M, Lopipero P, Blumenthal M, Ames BN (1987) Second chronological supplement to the carcinogenic potency database: standardized results of animal bioassays published through December 1984 and by the National Toxicology Program through May 1986. Environ Health Perspect 74: 237–329

Green MHL (1988) Short-term tests and the myth of the non-clastogenic mutagen. Mutagenesis 3: 369–371

Haworth S, Lawlor T, Mortelmans K, Speck W, Zeiger E (1983) *Salmonella* mutagenicity test results for 250 chemicals. Environ Mutagen 5 (Suppl 1): 3–142

Heddle JA (1988) Prediction of chemical carcinogenicity from in vitro genetic toxicity. Mutagenesis 3: 287–291

Howard PC, McCoy EC, Rosenkranz HS (1987) Sequential and differing nitroreductive pathways for mutagenic nitropyrenes in *Salmonella typhimurium*. Mutagenesis 2: 431–432

IARC (1987) IARC Monographs on the evaluation of carcinogenic risk to humans. Overall evaluations of carcionogenicity: an updating of IARC monographs volumes 1 to 42 (Suppl 7). International Agency for Research on Cancer. Lyon, France

ICPEMC (International Commission for Protection against Environmental Mutagens and Carcinogens) (1988) Testing for mutagens and carcinogens; the role of short-term genotoxicity assays. Mutation Res 205: 3–12

Karstadt M (1988) Quantitative risk assessment: qualms and questions. Teratogen Carcinogen Mutagen: 8: 137–152

Kier LD, Brusick DJ, Auletta AE, Von Halle E, Brown MM, Simmon VF, Dunkel V, McCann J, Mortelmans K, Prival M, Rao TK, Ray V (1986) The *Salmonella typhimurium*/mammalian microsomal assay. Mutation Res 168: 69–240

Lave LB, Omenn GS (1986) Cost-effectiveness of short-term tests for carcinogenicity. Nature 324: 29–34

Lave LB, Ennever FK, Rosenkranz HS, Omenn GS (1988) Information value of the rodent bioassay. Nature 336: 631–633

Matsushima T (1987) Chemical safety evaluation in Japan. 175–178

McCoy EC, Rosenkranz EJ, Mermelstein R, Rosenkranz HS (1983) Mutagenic specificity and the prediction of mechanisms and bioactivation pathways of genotoxicants: the mutagenicity of 5-nitroacenapthene as an example. Mutation Res 111: 61–68

Moore JA (1988) Risk assessment reappraisals: letter to the editor. Science 240: 1125

Mortelmans K, Haworth S, Lawlor T, Speck WT, Tainer B, Zeiger E (1986) *Salmonella* mutagenicity tests. II. Results from the testing of 270 chemicals. Environ Mutagen 8: 1–119

NAS (National Academy of Sciences) (1984) Toxicity testing: strategies to determine needs and priorities. National Academy Press, Washington DC

Perera FP (1984) The genotoxic/epigenetic distinction: relevance to cancer policy. Environ Res 34: 175–191

Perera F (1988) EPA cancer risk assessments: letter to the editor. Science 239: 1227

Reynolds SH, Stowers SJ, Patterson RM, Maronpot RR, Aaronson SA, Anderson MW (1987) Activated oncogenes in B6C3F1 mouse liver tumors: implications for risk assessment. Science 237: 1309–1316

Rosenkranz HS, Ennever FK (1988) Quantifying genotoxicity and nongenotoxicity. Mutation Res 205: 59–67

Shelby MD (1988) The genetic toxicity of human carcinogens and its implications. Mutation Res 204: 3–15

Stowers SJ, Maronpot RR, Reynolds SH, Anderson MW (1987) The role of oncogenes in chemical carcinogenesis. Environ Health Perspect 75: 81–86

Tennant RW, Margolin BH, Shelby MD, Zeiger E, Haseman JK, Spalding J, Caspary W, Resnick M, Stasiewicz S, Anderson B, Minor R (1987) Prediction of chemical carcinogenicity in rodents from in vitro genetic toxicity assays. Science 236: 933–941

Wilson JD (1988) Risk assessment reappraisals: letter to the editor. Science 240: 1126

Zeiger E, Anderson B, Haworth S, Lawlor T, Mortelmans K, Speck W (1987) *Salmonella* mutagenicity tests: III. Results from 255 chemicals. Environ Mutagen 9 (Suppl 9): 1–109

Zeiger E, Anderson B, Haworth S, Lawlor T, Mortelmans K (1988) *Salmonella* mutagenicity tests: IV. Results from the testing of 300 chemicals. Environ Mol Mutagen 11 (Suppl 12): 1–157

Subject Index

MIX
Papier aus verantwortungsvollen Quellen
Paper from responsible sources
FSC® C105338

If you have any concerns about our products,
you can contact us on
ProductSafety@springernature.com

In case Publisher is established outside the EU,
the EU authorized representative is:
Springer Nature Customer Service Center GmbH
Europaplatz 3, 69115 Heidelberg, Germany

Printed by Libri Plureos GmbH
in Hamburg, Germany